SHALL WE BELIEVE THE SCIENTISTS?

Do Scientists Believe in God?

StarTracker Publishing LLC

Silver Spring, MD 20910
www.startrackerpublishing.com

Paperback ISBN: 979-8-88760-001-7

According to the online Encyclopedia Britannica, "Venn diagram, graphical method of representing categorical propositions and testing the validity of categorical syllogisms, devised by the English logician and philosopher John Venn (1834–1923). Long recognized for their pedagogical value, Venn diagrams have been a standard part of the curriculum of introductory logic since the mid-20th century. Three-circle diagrams, in which each circle intersects the other two, are used to represent categorical syllogisms, a form of deductive argument consisting of two categorical premises and a categorical conclusion. A common practice is to label the circles with capital (and, if necessary, also lowercase) letters corresponding to the subject term of the conclusion, the predicate term of the conclusion, and the middle term, which appears once in each premise. If, after both premises are diagrammed (the universal premise first, if both are not universal), the conclusion is also represented, the syllogism is valid; i.e., its conclusion follows necessarily from its premises. If not, it is invalid."

SHALL WE BELIEVE THE SCIENTISTS?

Do Scientists Believe in God?

Dr. Garfield Greene &
David E. Greene, MBA, MBPA

This book is dedicated to our beloved sister,

the late Kathleen Mullins,

who inspired and encouraged us to complete this project.

A NOTE FROM THE PUBLISHER...

Welcome to StarTracker Publishing, where we believe in creating value for our readers.

Our team of experts is committed to delivering high-quality books that cater to the needs of our audience. We understand that today's readers are looking for books that are not only informative but also engaging and entertaining. That's why we strive to provide books that are compelling, persuasive, catchy, thought-provoking, and highly influential.

Our latest book is a must-read for everyone. We firmly believe that this book has the potential to leave a lasting impact and impression on its readers.

In our quest to bring you a rich and enlightening narrative, we delve into the lives and contributions of remarkable scientists spanning generations. From Aristotle and Einstein to Marie Curie and Francis Collins, the tapestry of scientific minds unfolds before you. Each chapter reveals the intricate threads of knowledge and faith woven into the fabric of their lives.

As you turn the pages of this book, you will not only gain valuable insights but also be inspired by the diverse perspectives on the intricate dance between science and faith.

We promise that the time and effort invested in reading, *Shall We Believe the Scientists? Do Scientists Believe in God?*, will yield exceptional insights, knowledge and pleasure beyond measure.

So, join us in this exploration of the profound connections between science and faith.

Your journey awaits, promising a transformative experience. This book invites you to explore the profound intersection of science and faith, offering a unique lens through which to view the complexities of life.

Happy Reading,

David Greene & Melissa Van Oss
StarTracker Publishing
https://startrackerpublishing.com/

PREFACE

This book is about science and faith. It addresses the role of faith in the lives of scientists, past and present, and the role of science in the lives of people of faith.

Although we are Christians, writing from a Judeo-Christian perspective, we do acknowledge the legitimacy of other points of view. We have a deep respect and appreciation for ancient and modern sciences. We believe that global warming is real, and a threat to life on earth and beyond, and, therefore, we have environmental responsibilities.

We believe in science, and we believe in God.

In our research for this book we have looked at the lives and contributions of a variety of scientists such as: Aristotle, Francis Bacon, Robert Boyle, Francis Collins, René Descartes, Thomas Edison, Michael Faraday, Alexander Fleming, and others. As you read this book, you will discover that the knowledge and pleasure you derive will be well worth your time and effort.

TABLE OF CONTENTS

INTRODUCTION

Scientists and religious leaders would agree that there is a difference between science and religion, and that one offers a different human experience than the other. Science is not religion, and religion is not science.

We have included a limited definition of some specific terms or categories used in this book in order to help the readers understand the information provided, and to help them to derive the most benefit from their reading. The following definitions can be expansive, but are brief here, as brevity is sufficient for our purpose.

Science

According to Encyclopedia Britannica, science is any system of knowledge that is concerned with the physical world and its phenomena and that entails unbiased observations and systematic experimentation. In general, science involves a pursuit of knowledge covering general truths or the operations of fundamental laws.

According to the editors of Encyclopedia Britannica, science can be divided into different branches, based on the subject of study. Based on an article recently revised and updated by *Erik Gregersen*, the physical sciences study the inorganic world and comprise the fields of astronomy, physics, chemistry, and the earth sciences. The biological sciences, such as biology and medicine, study the organic world of life and its processes; social sciences, such as anthropology and economics, study the social and cultural aspects of human behavior.

Religion

Religion, according to the editors of Encyclopedia Britannica, can be defined, in part, as human beings' relation to that which they regard as holy, sacred, absolute, spiritual, divine, or worthy of especial reverence. It is also commonly regarded as consisting of the way people deal with ultimate concerns about their lives and their fate after death. In many traditions these concerns are expressed in terms of one's relationship with or attitude toward gods or spirits; in the more humanistic or naturalistic forms of religion, they are expressed in terms of one's relationship with or attitudes toward the broader human community or natural world.[1]

Faith

Faith, in part, as defined by Encyclopedia Britannica is the inner attitude, conviction, or trust relating human beings to

---

[1] Editors of Encyclopedia Britannica

a supreme God or ultimate salvation. In religious traditions stressing divine grace, it is the inner certainty of an attitude of love granted by God himself. In Christian theology, faith is the divinely inspired human response to God's historical revelation through Jesus Christ and, consequently, is of crucial significance. No definition allows for the identification of "faith" with "religion."[2]

Judeo-Christian

According to Wikipedia, the free encyclopedia, the term Judeo-Christian is used to group Christianity and Judaism together, either in reference to Christianity's derivation from Judaism, Christianity's borrowing of Jewish scriptures to constitute the Old Testament of the Christian Bible, or due to the parallels or commonalities in Judeo-Christian ethics shared by the two religions. The term "Judeo-Christian" first appeared in the nineteenth century as a word for Jewish converts to Christianity. But Judaism is not Christianity, and Christianity is not Judaism.

Theology

The editors of the Encyclopedia Britannica define theology as the study of the nature of God and the relationship of the human and the divine. In this brief definition they say that the term was first used in the works of Plato and other Greek philosophers as they referred to the teaching of myth, but that the discipline expanded within Christianity and has found

---

[2] Encyclopedia Britannica

application in all theistic religions. Theistic religions hold to the view that all limited or finite things are in some way dependent on one supreme or one ultimate reality. One may ultimately speak of that reality in personal terms. In Judaism, Christianity, and Islam, that reality is often referred to as God.

Theology examines doctrines on such subjects as sin, faith, and grace, and considers the terms of God's covenant with humankind in matters such as salvation and eschatology. Eschatology concerns expectations of the end of the present age, human history, or of the world itself. The end of the world or end times is predicted by several World religions (both Abrahamic and non-Abrahamic), which teach that the negative world events will reach a climax. In further defining theology, it typically takes for granted the authority of a religious teacher or the validity of a religious experience. Theology differs from faith in that it is concerned with justifying and explaining a faith, rather than questioning the underlying assumptions of such faith, but it often employs quasi-philosophical methods.[3]

Theological Perspective

A person's theological perspective, for our purpose, is their way of thinking about the nature of God, faith, and religion.

---

[3] Encyclopedia Britannica

CHAPTER 1

Aristotle, Bacon, and Boyle

We begin with Aristotle because he offered something to the world of science; something that was unique. It was Aristotle that first defined and conducted empirical studies, that is, studies with concrete evidence, purpose, and logic. He was one of the first to offer the idea that reasoning and science are not mutually exclusive, and that they remain closely connected when conducting accurate and reliable studies. The scientific method permeates nearly every aspect of academic search and curiosity in general, and it is because of Aristotle's adherence to the logic of identifying problems and solving them that we are now able to conduct research.[4]

---

[4] Sites.middlebury.edu/fse1229pisapati/the-scientific-method/

ARISTOTLE

Aristotle was an ancient Greek philosopher and scientist who, along with Socrates and Plato, laid much of the foundation for western philosophy. Aristotle (c. 384 BC to 322 BC) is still considered one of the greatest thinkers in politics, psychology, and ethics.[5]

At the age of seventeen, he was sent to Athens to pursue a higher education. In Athens he enrolled in Plato's Academy, a preeminent place of learning in the Greek world, where he proved to be an exemplary scholar. He studied there for twenty years before he founded his own school, the Lyceum. Aristotle wrote approximately 200 works, most in the form of notes and manuscript drafts, touching on reasoning, rhetoric, politics, ethics, science, and psychology. Those works consist of dialogues, records of scientific observations, and systematic works. Of Aristotle's works, only thirty-one are currently

---

[5] Stanford Encyclopedia of Philosophy, first published Sat. March 18, 2000, revised Nov. 22nd 2022

in circulation, the majority of them dating to his time at the Lyceum.[6]

According to the Stanford Encyclopedia of Philosophy, syllogism had an immense influence on western thought. A syllogism is a form of deductive reasoning where one arrives at a specific conclusion or inference by examining ideas. An example would be: All violets are flowers. This is a violet. I'm holding a flower.

Did Aristotle believe in God?

The method that Aristotle used in an effort to show that the universe is a single causal system is through the examination of movement, which finds its culmination in book IX of the *Metaphysics*. Motion, according to Aristotle, refers to change in any of several different categories. According to Aristotle's fundamental principle, everything that is in motion is moved by something else. He says that there cannot be an infinite series of moved movers. He says that, "If it is true that when A is in motion there must be some B that moves A, then if B itself is in motion, there must be some C moving B, and so on. This series cannot go on forever, and so it must come to a halt in some X that is a cause of motion but does not move itself—an unmoved mover." Aristotle is willing to call the unmoved mover God.[7]

---

[6] En.wikipedia.org/wiki/Platonic Academy
[7] Ibid.

There are some biblical passages that would support what seems to me as Aristotle's theological perspective: *"In the beginning God created the heavens and the earth. Now the earth was formless and empty, darkness was over the surface of the deep, and the spirit of God was hovering over the waters. And God said, 'Let there be light, and there was light.'"* [8]

So even before there was a sun and a moon, there was light, and that light was created by the very word of God. *"Before the mountains were born you brought forth the whole world, from everlasting to everlasting you are God."* [9]

"I, the LORD, do not change. So you, the descendants of Jacob, are not destroyed." [10]

"Through him all things were made; without him nothing was made that has been made." [11] This New Testament passage is Christocentric, focusing on Jesus Christ, the second person of the Christian Trinity, and his role in creation as the Son of God.

---

[8] Genesis 1:1-3 NIV Bible
[9] Psalm 90:2 NIV Bible
[10] Malachi 3:6 NIV Bible
[11] John 1:3 NIV Bible

FRANCIS BACON

Sir Francis Bacon is important because he vigorously promoted the scientific method. He argued with those who believed in Aristotle's method. Instead, he advocated a lot of experimentation to prove things. Therefore, he is called the "father of empiricism."[12]

On January 2, 1561, Francis Bacon was born into a prominent, wealthy family in London, England. His father held a powerful position as Lord Keeper of the Great Seal, charged with the physical custody of the Great Seal of England.[13]

"Sir Francis Bacon provided readers with the first formal explanation of what we know today as the scientific method based on empirical evidence and inductive reasoning, which

---

[12] www.geni.com/people/Sir-Francis-Bacon-1st-and-last-Viscount-St-Alban-s/6000000006444780288

[13] Urbach, Peter Michael, Quinton, Anthony M., Quinton, Baron and Lea, Kathleen Marguerite. "Francis Bacon". *Encyclopedia Britannica*, 18 Jan. 2023, https://www.britannica.com/biography/Francis-Bacon-Viscount-Saint-Alban. Accessed 23 March 2023.

was solidified as the dominant scientific method through the works of great scientists such as Galileo and Isaac Newton."[14]

The *scientific method* is an ordered process that is used to establish scientific knowledge or change existing knowledge. It is accepted by many scientists and historians today as the foundation of modern science. The method usually contains the following steps:

1. Make an observation
2. Ask a question
3. Research the question
4. Propose a hypothesis
5. Test the hypothesis with an experiment
6. Draw a conclusion based on the experiment
7. Repeat the process[15]

According to the April 2023 issue of *Christianity Today*, Francis Bacon was a devout Anglican Christian who is noted for his public failure and his great scientific mind. He is quoted as saying, in part, "Perhaps the best of men are like the best of precious stones, wherein every flaw is noted more than those that are generally foul and corrupted."

Bacon introduced the essay form to the English language and completed his writing of *The New Atlantis*, which mixed his scientific approach and his Christian beliefs. He divided knowledge into philosophy, or natural knowledge, and divinity,

[14] Ibid.
[15] En.wikipedia.org/wiki/scientific method

or inspired revelation. Although he insisted that philosophy and the natural world needed to be studied inductively, he insisted that where religion is concerned, we can only study arguments for the existence of God. "Knowledge of God's nature, action and purposes," he said, "can only come from special revelation." He said that true study would ultimately help mankind. He said, "A little philosophy inclineth man's mind to atheism, but depth in philosophy bringeth men's minds about to religion." In other words, a little knowledge drives a person away from God, but deep study will bring them back.

It reminds me of a statement I read while in undergraduate school: *Little learning is a dangerous thing.* Little learning is a dangerous thing; knowing a little may make one mistakenly assume that one knows everything. The expression is a direct quotation from Alexander Pope's "Essay on Criticism" (1709), which echoed a sentiment stated in the sixteenth century by the French essayist Montaigne.[16]

In 1626 Francis stopped in the snow to conduct an experiment on the preservation of food, fell ill, and died on Easter Sunday. He included in his will this prayer:

"When I thought most of peace and honor thy hand was heavy on me, and hath humbled me, according to thy former loving-kindness...Just are thy judgements upon my sins...Be merciful unto me for my savior's sake, and receive me into thy bosom."[17]

---

[16] idioms.thefreedictionary.com/a little learning is a dangerous thing
[17] Https://www.fmousscientists.org/bacon What Was Robert Boyle's Contribution to the Atomic Theory?

Did Francis Bacon believe in God?

Bacon, in my opinion, is an example of a true scientist and a true Christian believer.

ROBERT BOYLE

Robert Boyle is largely regarded today as the first modern chemist and, therefore, one of the founders of modern chemistry as well as one of the pioneers of modern experimental scientific method.[18]

Robert Boyle was an Anglo-Irish philosopher and writer who was born on January 25, 1627, in Lismore Castle, County Waterford, Ireland, and died on December 31, 1691, in London, England. He was a philosopher and theological writer. Boyle was also a preeminent figure of the seventeenth

---

[18] En.wikipedia.org/wiki/Robert_Boyle

century intellectual culture. Robert Boyle was particularly known in the field of chemistry. However, his scientific work covered many other areas, which included hydrostatics, physics, medicine, earth sciences, natural history, and alchemy.[19]

The dictionary defines alchemy as the medical forerunner of chemistry, based on the supposed transformation of matter. It was concerned particularly with the attempts to convert base metals into gold or to find a universal elixir. He also produced Christian devotional and ethical essays as well as tracts on biblical language, the limits of reason, and the role of the natural philosopher as a Christian.

He discovered that the volume of a gas decreases with increasing pressure and vice versa, which is known as Boyle's law. Robert Boyle is known as the "Father of Chemistry" due to his discovery that atoms must exist, based on the relationship between pressure and volume of gas. His theorem called Boyle's Law reasons that because a fixed mass of gas can be compressed, gas must be made of particles, or atoms, because there is space between them.[20]

Boyle was a devout Anglican, and with the rise of science and reason during his lifetime, he was troubled by increasing atheism. This encouraged him to write about his belief and religion supporting each other.

---

[19] www.sciencehistory.org/historical-profile/robert-boyle
[20] Ibid.

Did Robert Boyle believe in God?

Robert Boyle is an outstanding example of a Christian scientist whose faith interacted fundamentally with his science. His remarkable piety was the driving force behind his interest in science and his Christian character shaped the ways in which he conducted his scientific life. A deep love for scripture, coupled, ironically, with a lifelong struggle with religious doubt, led him to write several important books relating to scientific and religious knowledge. Ultimately, he was attracted to the mechanical philosophy because he thought it was theologically superior to traditional Aristotelian natural philosophy; by denying the existence of a quasi-divine "nature" that functioned as an intermediary between God and the world, it more clearly preserved God's sovereignty and more powerfully motivated people to worship their Creator.[21]

Robert Boyle was a brilliant scientist who contributed a great deal to modern science and wrote extensively about faith in God and the necessary relationship between faith and science.

---

[21] Robert Boyle's Religious Life, Attitudes, and Vocation, Science Christian Belief, June 2007.https://www.scienceanchristianbelief.org/view abstract.hp?ID=917

CHAPTER 2

Descartes, Edison, and Collins

Descartes is regarded as one of the greatest philosophers in history. He was responsible for a tremendous conceptual breakthrough through his analytical geometry, linking the previously separated fields of geometry and algebra. He proved that he could solve previously unsolvable problems in geometry by converting them to simple problems in algebra.[22]

René Descartes was born on March 31, 1596, in France and died on February 11, 1650, in Stockholm, Sweden. He was a French math-ematician, scientist, and philosopher. He was one of the first to abandon *Scholastic Aristotelianism.* He also formulated the first modern version of the *mind-body dualism* from which stems the mind-body problem. He also supported the development of a *new science* grounded in observation and experiment. Because of those innovations, Descartes is generally regarded as the founder of modern philosophy.[23] Descartes applied an original system of methodical doubt. He dismissed apparent knowledge coming from authority, the senses, and reason and created new ways of thinking about knowledge, based on his intuition that when he thinks, he exists. This he expressed in the proclamation, "I think, therefore I am." Best known in the Latin translation, *"cogito, ergo sum,"* originally written in French, *"Je pense, donc, je suis."*[24]

He is also credited with developing *Cartesian Dualism* which is also referred to as *mind-body dualism*, which is based on the metaphysical argument that the mind and the body are two different substances that interact with one another.[25] His

---

[22] www.famousscientists.org/rene-descartes/
[23] Watson, Richard A.. "René Descartes". *Encyclopedia Britannica*, 27 Feb. 2023, https://www.britannica.com/biography/Rene-Descartes. Accessed 24 March 2023.
[24] Ibid.
[25] Watson, Richard A.. "René Descartes." *Encyclopedia Britannica*, 27 Feb. 2023, https://www.britannica.com/biography/Rene-Descartes. Accessed 24 March 2023.

primary contribution to the field of mathematics came from bridging the gap between algebra and geometry, resulting in the Cartesian coordinate system which is still in use today.[26]

As he continued his reasoning process, Descartes examined the philosophical possibility of the existence of God, in his *Third Meditation*. He divided this evidence into two umbrella categories, which he called proofs. His logic is not difficult to follow.

In the first proof, Descartes argues that, by evidence, he is an imperfect being who has an objective reality, including the notion that perfection exists, and therefore has a distinct idea of a perfect being (God, for example).

Further, Descartes realizes that he is less formally real than the objective reality of perfection, and therefore there has to be a perfect being existing formally from whom his innate idea of a perfect being derives, wherein he could have created the ideas of all substances, but not the one of God.[27]

There had to be a perfect being, namely God, in order to give him the idea of a perfect God.

The second proof then goes on to question who it is then that keeps him—having an idea of a perfect being—in

[26] Stanford Encyclopedia of Philosophy
[27] Borghini, Andrea. "Rene Descartes' "Proofs of God's Existence","ThoughtCo, Aug 27,2020, thoughtco.com/Descartes-3-proofs-of-gods-existence-2670585

existence, eliminating the possibility that he himself would be able to do it. He proves this by saying that he would owe it to himself, if he were his own existence maker, to have given himself all sorts of perfections. The very fact that he is not perfect means that he would not bear his own existence. Similarly, his parents, who are also imperfect beings, could not be the cause of his existence since they could not have created the idea of perfection within him. That leaves only a perfect being, God, that would have had to exist to create and be constantly recreating him.[28]

Did René Descarte believe in God?

Essentially Descartes's proofs rely on the belief that by existing, and being born an imperfect being (but with a soul or spirit), one must, therefore, accept that something of more formal reality than ourselves must have created us. Basically, because we exist and are able to think ideas, something must have created us.[29]

So, as I understand him, Descartes is saying that we understand God as a perfect being. He is saying that it is more perfect to exist than not to exist. Therefore, God exists.

---

[28] Ibid.
[29] Ibid.

THOMAS EDISON

Thomas Edison is considered one of America's leading businessmen and innovators. He rose from humble beginnings to work as an inventor of major technology. He invented the first commercially viable incandescent light bulb, and is credited today for helping to build America's economy during the industrial revolution.[30]

Thomas Alva Edison was an American inventor who singularly or jointly held the world record of 1,093 patents. He also created the first industrial research laboratory in the world.[31] Before his death in 1931, he had been instrumental in introducing the modern age of electricity. From his laboratories and his workshops we have received the phonograph, the carbon-button transmitter for the telephone speaker and microphone, the first commercial electric light and power system, as

---

[30] www.biography.com/inventor/thomas-edison

[31] Josephson, Matthew and Conot, Robert E.. "Thomas Edison." *Encyclopedia Britannica*, 7 Feb. 2023, https://www.britannica.com/biography/Thomas-Edison. Accessed 24 March 2023.

well as an experimental electric railroad, and the most important elements of the motion picture apparatus.[32]

Edison's phonograph was quite innovative, but people saw it primarily as a novelty. He had moved on to another world-altering concept which was the incandescent light bulb.[33]

Electric light bulbs had been in existence since the early nineteenth century, but they were not very durable due to their filaments. Filaments form the part that produces light. An early form of electric light, the carbon arc light relied on the vapor of battery-heated carbon rods to produce light. Those, however, had to be lit by hand, and the bulb was short-lived and not reliable. They were too expensive and not practical.

Edison's, by contrast, were cheap, practical, and lasted much longer. After years of dedicated work on improving the light bulb, Edison made filaments or threads of many heat-resisting materials into glass globes. He continued his research for many months, and, in October 1879, he introduced the modern age of light. He invented the first commercially practical incandescent electric light. He also improved the dynamo to furnish the power needed for electric lighting systems. Edison also developed a complete system of distributing the current, and in 1882, built the first central power station in lower Manhattan.[34]

---

[32] Ibid.

[33] Thomas Edison didn't invent the light bulb—but here's what he did do (nationalgeographic.com)

[34] Ibid.

Did Thomas Edison believe in God?

In an interview in the *New York Times* magazine, Mr. Edison said, "I do not believe in the God of the theologians, but that there is a supreme intelligence I do not doubt." He also said, "Nature made us—nature did it all—not the gods of the religions." Still, Edison insisted that the statement did not mean that he was an atheist. Edison has been called a deist, a pantheist, and a free thinker. In general, Deism refers to what can be called *natural religion*, the acceptance of a certain body of religious knowledge that is inborn in every person or that can be acquired by the use of reason, and the rejection of religious knowledge when it is acquired through revelation or teaching of any church.[35]

It does not appear that Edison called himself a pantheist, but some people apparently thought he was. Pantheism is the belief that God and the universe are the same. A pantheist believes, in part, that everything that exists is a part of God or that God is a part of everything that exists. The name comes from the Greek words "theism," meaning belief in God, and "pan," meaning all. Panentheism is similar but not the same as pantheism. Panentheism is the belief that the divine interpenetrates every part of the universe, and also extends beyond space and time.[36]

---

[35] Pailin, David A. and Manuel, Frank Edward. "Deism". Encyclopedia Britannica, 17 Feb. 2023, https://www.britannica.com/top/Deism. Accessed 2 March 2023. www.thetimesherald.com/story/life/faith/2017/02/10/Edisons-spirituality

[36] Simple.wikipedia.org/wiki/Pantheism

Did Thomas Edison believe in God?

Yes he did, but it is evident by his statements that he did not believe in a personal god. However, it is not so difficult to conceive of such a belief. As a child, I believed in God as the Creator, but not as a personal God, and certainly not as a personal Savior.

FRANCIS COLLINS

Francis Collins is an American geneticist. He discovered genes that cause genetic diseases. He was the director of the United States National Institutes of Health (NIH) from 2009 to 2021.[37]

Francis Collins, PhD, MD, was homeschooled by his mother for much of his youth. As a child, he took an interest in science. In 1970 he received a bachelor's degree from the

---

[37] www.britannica.com/biography/Francis-Collins

University of Virginia, and, subsequently, degrees from Yale University and the University of North Carolina at Chapel Hill. In 1984 Dr. Collins joined the staff of the University of Michigan at Ann Arbor as an assistant professor. There he earned the reputation as one of the world's foremost genetic researchers.

In 1989 Collins announced the discovery of the gene that causes cystic fibrosis, an inherited life-threatening disorder that harms the lungs and digestive system. In the following year, a team led by Dr. Collins found the gene that causes neurofibromatosis, a genetic disorder that generates the growth of tumors.

As the director of the human Genome project Dr. Collins oversaw a fifteen-year endeavor to completely map the sequence of human DNA by 2005. According to many scientists and medical researchers, this was the most important scientific endeavor in modern times.[38]

According to *Oxford Languages*, DNA, *deoxyribonucleic acid*, is a biochemical term. It is a self-replicating material that is present in nearly every living organism as the main constituent of chromosomes, and is the carrier of genetic information. It can be alternatively defined as the fundamental and distinctive characteristics or qualities of someone or something, especially when regarded as unchangeable.

In his book *Organic Church*, Neil Cole defines the DNA of a church as divine truth, nurturing relationships, and apostolic

---

[38] Craine, Anthony G. "Francis Collins". *Encyclopedia Britannica*, 1 Mar. 2023. https://www.britannica.com/biography/Francis-Collins. Accessed 16 March 2023

mission. He says that these three strands give a church its biblical identity as expressed in the Scriptures and determine their validity and their ability to replicate and multiply what matters.[39]

According to Eric Green, MD, PhD, the genome is the entire set of DNA instructions found in a cell. "In humans," he says, "the genome consists of twenty-three pairs of chromosomes, located in the cell's nucleus, as well as a small chromosome in the cell's mitochondria." He says that a genome contains all the information needed for an individual to develop and function.[40]

Does Francis Collins believe in God?

In his book, *The Language of God*, published by Simon and Schuster, Dr. Collins explains in detail why he believes that faith in God and faith in science can coexist within a person and be harmonious. He has heard the arguments against faith by some scientists, and refutes them, and the arguments against science by some people of faith, and can counter those arguments as well. He tells of his journey from atheism to faith, and takes the readers to an interesting tour of modern science where he proves that physics, chemistry, and biology can fit together with belief in God and in the Bible.[41]

---

[39] www.compellingtruth.org/organic-church.html
[40] www.genome.gov ›eric green Accessed Oct 19, 2022
[41] Craine, Anthony G.. "Francis Collins". *Encyclopedia Britannica*, 1 Mar. 2023, https://www.britannica.com/biography/Francis-Collins. Accessed 27 March 2023.

Dr. Collins grew as an agnostic, believing that the existence of God was unknowable. Later he became an atheist, not believing in God at all, during his PhD in chemistry. According to his personal testimony as recorded in his book, *The Language of God*, the harsh reality that he faced as he watched patients die led him to question his own religious views, and at the age of twenty-seven he became a Christian. Francis Collins is a top scientist and an Evangelical Christian. Collins was deeply involved in developing and testing the vaccines currently in use in the United States against COVID. He said that a lot of prayer as well as a lot of science had gone into the research. He believed that those prayers were answered. He said that it felt like a gift from God, but that you have to unwrap that gift. "Give God the glory for the COVID vaccines but roll up your sleeves."[42]

It appears to me that, in Dr. Collins's mind, we can accept and benefit from some of the gifts brought forth by science, especially in the field of medicine. At the same time, he seemed to agree that we should be grateful to our Creator, since God is the one that is ultimately responsible for those gifts. I believe that an applicable New Testament scripture here might come from James 1:17: *"Every good and perfect gift is from above, coming down from the Father of heavenly lights, who does not change like shifting shadows."*[43]

---

[42] Ibid.
[43] Holy Bible, New international Version

I have rolled up my sleeves a few times recently, and I always pray as I am doing so. I believe in the vaccines, and I believe in God who has given the knowledge and the means to develop them.

CHAPTER 3

Faraday, Fleming, and Fibonacci

Faraday, an English physicist and chemist was one of the most predominant scientists of the nineteenth century. His many experiments have advanced the understanding of electromagnetism. In 1820 he produced the first known compounds of carbon and chlorine. In 1821 Michael Faraday invented the first electric motor. In addition, in the 1830s he discovered a method for converting mechanical energy into electricity on a large scale, creating the electric generator.[44]

Electromagnetism is used to understand a number of phenomena. It is used to accelerate various charged particles which are then used in other applications. Electromagnetism is used in medical machines such as the MRI. The X-ray is based on electromagnetism, which holds an important place in the field of magnetism.[45]

One of the greatest scientists in history came from a very poor family. Michael Faraday lived from 1791 to 1867. This was a time when the study of science was usually limited to people born into wealthy families.

Faraday's Beginnings

Faraday was born on September 22, in London, England. His parents were James and Margaret. His father worked as a blacksmith, and suffered from poor health. Before she was married, his mother had been a servant. Michael, his mother and father, and his two older siblings lived in a degree of poverty.[46]

---

[44] Williams, L. Pearce. "Michael Faraday". *Encyclopedia Britannica*, 1 Apr. 2023, https://www.britannica.com/biography/Michael-Faraday. Accessed 4 April 2023..

[45] Kashy, Edwin, McGrayne, Sharon Bertsch and Robinson, Frank Neville H. "Electromagnetism." *Encyclopedia Britannica*, 15 Feb. 2023, https://www.britannica.com/science/electromagnetism. Accessed 4 April 2023.

[46] "Michael Faraday." Famous Scientists. famousscientists.org. 24 Nov. 2014. Web. 4/4/2023 <www.famousscientists.org/michael-faraday/>.

After receiving his basic education in a local school where he attended until the age of thirteen, Michael started working as a delivery boy for a bookshop in order to help support the family. After a year of hard work, his employer made him an apprentice bookbinder. As an avid reader, and eager to learn more about the world, Faraday did not restrict his reading to the books he found in the store. He also found himself reading more and more about science. He was particularly drawn to two important sources of knowledge. One was the *Encyclopedia Britannica*, from which he gained his electrical knowledge and more, and the other was a book by Jane Marcet, *Conversations on Chemistry*. It was 600 pages of chemistry for ordinary people.[47]

The following are some of Michael Faraday's most notable discoveries and achievements:

In 1821 he discovered electromagnetic rotation.

In 1823 he made important discoveries concerning gas liquefaction and refrigeration, providing hard evidence for Dalton's belief that all gasses could be liquified by the use of low temperatures and/or high pressures.

In 1825 he discovered Benzene, one of the most important substances in chemistry.

In 1831 he discovered electromagnetic induction.

In 1834 he discovered "Faraday's Laws of Electrolysis," one of the major contributors to the science of

---

[47] Marcet, Jane, 1883, *Conversations on Chemistry*, Oxford, England.

electrochemistry which produced Li-ion batteries and other batteries used in modern mobile technology.

In 1826 he invented the Faraday cage.[48]

A Faraday cage is a protective enclosure that prevents certain types of electromagnetic radiation from entering or exiting, according to the Florida State University Magnetic Field Laboratory.

We use Faraday cages on a pretty regular basis in places like hospitals and even in our kitchen. So a Faraday cage is essentially a container, or a shield, that blocks out electromagnetic radiation from across the electromagnetic spectrum, such as radio waves and microwaves, according to Florida State University.

It works on the principle that when an electromagnetic field hits something that can conduct electricity, the charges remain on the exterior of the conductor rather than traveling inside. In more practical terms, that means that a cage constructed of a material that can conduct electricity will prevent certain electromagnetic radiation from passing through. This applies to both constant, or static, electric fields, and changing, or non-static, electric fields.[49]

If you have a microwave in your[50] kitchen, that is a form of Faraday cage, because it keeps microwaves trapped inside the machine so that they heat your food and do not escape out.

---

[48] "Michael Faraday." Famous Scientists. famousscientists.org. 24 Nov. 2014. Web. 4/4/2023 <www.famousscientists.org/michael-faraday/>. Published by FamousScientists.org.
[49] Ibid.
[50] Ibid.

On a much grander scale, Magnetic Resonance Imaging (MRI) scanners in medical settings use Faraday cages to prevent radio signals from entering the room and making the equipment safe to use.[51]

Did Michael Faraday believe in God?

Michael Faraday was a seriously committed Christian. He was a member of a "fundamentalist," nonconformist group who called themselves the Sandemanians. They believed in simple, humble living, and caring for the disadvantaged in community, and the absolute truth of the gospel.[52] The Sandemanian sect was founded around 1730 in Scotland by John Glas, who lived from 1695 to 1773. Glas, who was a Presbyterian minister in the church of Scotland, concluded that there was no support in the New Testament for a national church because the kingdom of Christ was essentially spiritual. There was a strong link between the Sandemanians and scientists.[53]

---

[51] MRI Shielding | MRI Shielded Room | Faraday Cage | Faraday Cages - European EMC Products (euro-emc.co.uk)
[52] MRI Shielding | MRI Shielded Room | Faraday Cage | Faraday Cages - European/ www.christiantoday.com.au/news/michael-faraday-his-christian-faith-influenced-his-science.html
[53] www.britannica.com/topic/Sandemanians

ALEXANDER FLEMING

Sir Alexander Fleming lived from 1881 to 1955. He was a Scottish physician and microbiologist, best known for his discovery of the world's first broadly effective antibiotic substance, which he called penicillin.[54]

It was while serving in the Royal Army Medical Corps in World War I that he conducted research on antibacterial substances that would be non toxic to human beings. In 1928 he inadvertently made the discovery when he noticed that a mold contaminating a bacterial culture was inhibiting the bacteria's growth. Fleming shared a 1945 Nobel Prize with Ernst Boris Chain and Howard Walter Florey, who both carried Fleming's basic discovery further, isolating, purifying, testing, and producing penicillin in quantity.[55]

---

[54] en.wikipedia.org/wiki/Alexander_Fleming
[55] "Britannica, The Editors of Encyclopaedia. "Sir Alexander Fleming summary." *Encyclopedia Britannica*, 14 Oct. 2003, https://www.britannica.com/summary/Alexander-Fleming. Accessed 10 April 2023.

Flemings's most significant contributions to science were: proving that antiseptics kill rather than cure, the discovery of lysozyme, and the discovery of penicillin.

In 1914, when World War I broke out, Fleming was thirty-three years old. At that time he decided to join the army, and subsequently became a captain in the Royal Army Medical Corps, and worked in the field hospitals in France. While serving there he established, through a series of brilliant experiments, that the agents used to treat wounds and prevent infections were actually killing more soldiers than the infections were. He discovered that carbolic acid, boric acid, and hydrogen peroxide were not killing bacteria deep in the wounds. In fact, they were lowering the soldier's natural resistance to infection because they were killing white blood cells. Fleming demonstrated that the antiseptic agents were only useful in treating superficial wounds, but that they were harmful when applied to deep wounds.[56]

Amroth Wright believed that a saline solution—salt water—should be used to clean deep wounds since that process did not interfere with the body's own defenses, and attracted white cells. Fleming demonstrated and proved that process in the field. Wright and Fleming published those results, but most doctors refused to change their mode of treatment resulting in many preventable deaths.[57]

[56] "Alexander Fleming." Famous Scientists. famousscientists.org. 09 Jul. 2015. Web. 4/10/2023 <www.famousscientists.org/alexander-fleming/>. Published by FamousScientists.org.
[57] Ibid.

In 1919 he resumed his research work at St. Mary's Hospital Medical School in London. His wartime experience helped Fleming to establish his view that antibacterial agents should be used only if they worked with the body's natural defenses rather than against them; in particular, the agent must not harm white blood cells.

Fleming discovered such an agent first in 1922 when he was forty-one years old. He had taken secretions from inside the nose of a patient who had a head cold and cultured the secretions to grow any bacteria that happened to be present. In the cultures he found a new bacterium that he called *Micrococcus lysodeikticus*, now called M luteus.

A few days later when Fleming himself was suffering from a head cold, while examining the bacteria, a drop of mucus from his nose fell on to the bacteria. The bacteria in the area where the drop fell were almost instantly destroyed. Since Fleming was always on the lookout for bacteria killers that observation excited him tremendously.[58] Fleming tested the effect of other body fluids, such as blood serum, saliva, and tears on these bacteria and found that bacteria would not grow where a drop of one of these fluids was placed.[59]

He discovered the common factor in the fluids was a newly discovered enzyme named *lysozyme*. He discovered that lysozyme destroys certain types of microbes, rendering them harmless to people. When present in our bodies lysozyme prevents

---

[58] Alexander Fleming." Famous Scientists. famousscientists.org. 09 Jul. 2015. Web. 4/10/2023 <www.famousscientists.org/alexander-fleming/>.
[59] Ibid.

some potential pathogenic microbes from causing us harm, giving us natural immunity to a number of diseases. However, lysozyme is of little use as a medicine, since it has little or no effect on many microbes that infect humans.[60] Fleming had been successful in discovering a natural antibiotic that did not kill white blood cells. He needed to find a more powerful antibiotic. Then medicine could be transformed.

Today, lysozyme is used as a preservative for some food and wine. It is present in large amounts in egg whites, protecting chicks against infection. Lysozyme is also used in medicines, particularly in Asia, where it is used to treat head colds, athlete's foot, and throat infections.[61]

The bacteriologist Alexander Fleming said,

"The view has been generally held that the function of tears, saliva and sputum, so far as infections are concerned, was to rid the body of microbes, by mechanically washing them away…however, it is quite clear that these secretions, together with most of the tissues of the body, have the property of destroying microbes to a very high degree."

After spending a long vacation with his wife and son in August of 1928, Fleming returned to his laboratory. On Monday, September 3, he saw a pile of petri dishes he had left

---

[60] Ibid.
[61] "Alexander Fleming." Famous Scientists. famousscientists.org. 09 Jul. 2015. Web. 4/10/2023 <www.famousscientists.org/alexander-fleming/>.

on his bench. The dishes contained colonies of *staphylococcus* bacteria. In his absence one of his assistants had left a window open and one of the dishes had become contaminated by different microbes. (A microbe is a microorganism, especially a bacterium, that causes disease or fermentation.) Fleming was annoyed; and as he looked through the dishes, he found that something remarkable had taken place in one of them. A fungus was growing, and the bacterial colonies around it had been killed. Away from the fungus, the bacteria appeared to be normal. He was excited by his observation and showed the dish to his assistant who remarked about how similar this seemed to Fleming's famous discovery of lysozyme.

Fleming hoped he had discovered a better antibiotic than lysozyme. So he devoted himself to growing more of the fungus. He identified it as belonging to the *Penicillium* genus, and that it produced a bacteria-killing liquid. On March 7, 1929, he formally named the antibiotic penicillin. Fleming published his results, showing that penicillin killed many different types of bacteria, including those responsible for scarlet fever, pneumonia, meningitis, and diphtheria. In addition, penicillin was non-toxic (nonpoisonous) and it did not attack white blood cells.[62]

In his Nobel Prize winning speech in 1945, Alexander Fleming warned of a danger which today is becoming ever more pressing. He predicted antibiotic resistance. He said,

[62] Ibid.

"It is not difficult to make microbes resistant to penicillin in the laboratory by exposing them to concentrations not sufficient to kill them, and the same thing has occasionally happened in the body. The time may come when penicillin can be bought by anyone in the shops. Then there is the danger that the ignorant man may easily underdose himself, and by exposing his microbes to non-lethal quantities of the drug, make them resistant."

Fleming always had high praise for Florey, Chain, and their team, and he downplayed his own role in penicillin's story. In spite of his modesty, he became a worldwide hero, and millions of people owed their lives to the antibiotic that he had discovered.

In 1945 Fleming toured America where chemical companies offered him a personal gift of $100,000 as a mark of respect and gratitude for his work. True to his character, Fleming did not accept the gift for himself, but donated it to the research laboratories at St. Mary's Hospital Medical school.[63]

Did Alexander Fleming believe in God?

Alexander Fleming did believe in God, although he was not publicly very religious. He did, however, recognize the presence and influence of God in his work as a microbiologist. Consider the following abstract:

---

[63] Ibid.

In logic and reasoning, a signature indicates the presence of an author; likewise, the characteristics of staphylococci indicate the presence of a Creator. Staphylococci and its "kind" are common bacteria, particularly in colonized people. Staphylococcus aureus has a complex molecular mechanism of assembling its golden pigment, staphyloxanthin. The biosynthesis of staphyloxanthin is a stellar example of irreducible complexity. Similar to staphylococci, the life and works of Alexander Fleming show the fingerprints of Providence. The so-called "serendipitous" achievements of Fleming have contributed to modern medicine, convincing Fleming and others that God was at work in his life. Fleming recognized that his life's discoveries and the "weaving" of events were more than chance; it was the invisible hand of God in his life and works. The molecular complexities of staphylococci mechanisms and the achievements of Fleming indicate the signature of a divine Designer who has placed his signature on his art piece, staphylococci.[64]

---

[64] Gillen, Alan L. and Cargill, Michael, "The Signature of God in Medicine and Microbiology: An Apologetic Argument for Declarative Design in the Discoveries of Alexander Fleming." (2016). *Faculty Publications and Presentations*. 132. https://digitalcommons.liberty.edu/bio_chem_fac_pubs/132

FIBONACCI

Leonardo Pisano Fibonacci was born in 1170 and died around 1240 or 1250. He was an Italian number theorist. Fibonacci introduced the world to such wide-ranging mathematical concepts as what we now call the Arabic numbering system, the concept of square roots, number sequencing, and math word problems. He is especially known for the Fibonacci sequence.[65]

Although Fibonacci was born in Italy, he obtained his education in North Africa. Very little is known about him or his family, and there are no photographs or drawings of him. Much of the information about Fibonacci has been gathered by his autobiographical notes, which he included in his books.

---

[65] "Leonardo Pisano Fibonacci." *Fibonacci* (1170-1250), History.mcs.st-andrews.ac.uk

Mathematical Contributions

Fibonacci is considered to be one of the most talented mathematicians of the Middle Ages. Few people realize that it was Fibonacci that gave the world the decimal number system (Hindu-Arabic numbering system), which replaced the Roman numeral system. When he was studying mathematics, he used the Hindu-Arabic (0-9) symbols instead of Roman symbols, which didn't have zeros and lacked place value. In fact, when using the Roman numeral system, an abacus was usually required. There is no doubt that Fibonacci saw the superiority of using Hindu-Arabic system over the Roman Numerals.[66]

Liber Abaci

Fibonacci showed the world how to use what is now our current numbering system in his book *Liber Abaci*, which he published in 1202. The title translates as "The Book of Calculation." The following problem was written in his book: "A certain man put a pair of rabbits in a place surrounded on all sides by a wall. How many pairs of rabbits can be produced from that pair in a year if it is supposed that every month each pair begets a new pair, which from the second month on becomes productive?"[67]

It was this problem that led Fibonacci to the introduction of the Fibonacci numbers and the Fibonacci sequence, which is what he remains famous for to this day. The sequence is 1, 1, 2,

---

[66] Knott, R. "Who was Fibonacci?" Maths.surrey.ac.uk.

[67] Russell, Deb. "Biography of Leonardo Pisano Fibonacci, Noted Italian Mathematician." ThoughtCo, Feb. 16, 2021, thoughtco.com/leonardo-pisano-fibonacci-biography-2312397.

3, 5, 8, 13, 21, 34, 55... This sequence shows that each number is the sum of the two preceding numbers. It is a sequence that is seen and used in many different areas of mathematics and science today. The sequence is an example of a recursive sequence.

The Fibonacci sequence defines the curvature of naturally occurring spirals, such as snail shells and even the pattern of seeds in flowering plants. The Fibonacci sequence was actually given the name by a French mathematician Edouard Lucas in the 1870s.

In addition to *Liber Abaci*, Fibonacci authored several other books on mathematical topics ranging from geometry to squaring numbers (multiplying numbers by themselves). The city of Pisa (technically a republic at that time) honored Fibonacci and granted him a salary in 1240 for his help in advising Pisa and its citizens on accounting issues. Fibonacci died between 1240 and 1250 in Pisa.

Fibonacci is famous for his contributions to number theory.[68]

In his book, *Liber Abaci*, he introduced the Hindu-Arabic place-value decimal system and the use of Arabic numerals into Europe.

He introduced the bar that is used for fractions today; previous to this, the numerator had quotations around it.

The square root notation is also a Fibonacci method.

It has been said that the Fibonacci numbers are nature's numbering system and that they apply to the growth of living

---

[68] Ibid.

things, including cells, petals on a flower, wheat, honeycomb, pine cones, and much more.

A notable quote from Fibonacci, "If by chance I have omitted anything more or less proper or necessary, I beg forgiveness, since there is no one who is without fault and circumspect in all matters."[69]

Did Fibonacci believe in God?

One quote that is attributed to Fibonacci is, "The fingerprints of God are all around us. Just take a look."[70] We find it in plants, animals, space, art, classical music, and much more. For many amazing illustrations you can Google *Fibonacci and the finger of God.*

[69] Russell, Deb. "Biography of Leonardo Pisano Fibonacci, Noted Italian Mathematician." ThoughtCo, Feb. 16, 2021, thoughtco.com/ leonardo-pisano-fiboonacci-biography-2312397.
[70] The Fingerprints Of God Are All Around Us. Just Take A Look! | FaithHub

CHAPTER 4

Carver, Einstein, and Ross

GEORGE WASHINGTON CARVER

George Washington Carver was born in 1861 near Diamond Grove, Missouri. He died in 1943 in Tuskegee, Alabama. He was an agricultural chemist and agronomist. Born into slavery, Carver lived until the age of ten or twelve on his former owner's plantation. Then he left and worked at various menial jobs. It was not until he was in his late twenties that he obtained a high school education. Afterward, he attended the Iowa State College School of Agriculture, where he obtained bachelor's and master's degrees. In 1896 he joined Booker T. Washington at the Tuskegee Institute, which is now Tuskegee University in Alabama. There he became the director of agricultural research.

Soon he began promoting the planting of peanuts and soybeans, which he knew would restore the fertility of the soil that had been depleted by the growth of cotton. He worked intensively with the sweet potato and the peanut, which was at that time not even recognized as a crop. Ultimately he developed 118 derivative products from sweet potatoes and 200 from peanuts. Carver's efforts helped to liberate the South from its untenable cotton dependency; and by 1940 the peanut was the South's second largest cash crop. During the Second World War, he devised 500 dyes to replace those that were no longer available from Europe. Despite his international acclaim and extraordinary job offers, Carver remained at Tuskegee for the rest of his life, and in 1940 he donated his life's savings to establish the Carver Research Foundation at Tuskegee.[71]

It is highly probable that no other scientist has had to face as many social barriers as George Washington Carver, the black American botanist noted for revolutionizing agriculture

---

[71] Britannica, The Editors of Encyclopaedia. "George Washington Carver summary." *Encyclopedia Britannica*, 14 Oct. 2003.

in the southern United States. He was born toward the end of the Civil War to a slave family on the farm of Moses Carver, a German American Settler. When he was just an infant, he and his mother and sister were kidnapped by Kentucky night raiders.

It is not known what happened to his mother and sister, but George was rescued and returned to the Carver family, who raised him and his brother James. He grew up in a strictly segregated south where very few black schools were available. However, George's strong desire for learning prompted him to persevere, and he earned his diploma from Minneapolis High School in Minneapolis, Kansas.

For him, entering college was very difficult. However, he was eventually accepted at Simpson College in Indianola, Iowa, to study art. In 1891, he transferred to Iowa State Agriculture College in Ames, which is now Iowa State University, to study botany. He was the first black student and, later, the first black faculty member there. While he was there, he adopted the middle name "Washington" to distinguish himself from another George Carver. He received an undergraduate degree in 1884, and his master's degree in 1896. He became a nationally recognized botanist for his work in plant pathology, and mycology (the study of fungi). He joined Booker T. Washington at the Tuskegee Normal and Industrial Institute in Alabama to teach former slaves how to farm for self-sufficiency. He conducted numerous research projects that also contributed to medicine and other fields, and used his influence to champion the relief of racial tensions.

Carver said, "One reason I never patent my products is that if I did, it would take so much time, and I would get nothing else done. But mainly I don't want my discoveries to benefit specific favored persons."[72]

Did George Washington Carver believe in God?

Carver was frugal in finance, humble in character, and undoubtedly a deeply devoted Christian. He attributed inspiration of his work to God.[73] His studies of nature convinced him of the existence and benevolence of the Creator: "Never since have I been without this consciousness of the Creator speaking to me…the out of doors has been to me more and more a great cathedral in which God could be continuously spoken to and heard from."[74]

Carver died on January 5, 1943, of complications from injuries he incurred in a bad fall. His life savings of $60,000 was donated to the museum and foundation bearing his name. The epitaph on his grave on the Tuskegee University campus summarizes the life and character of this former slave, man of science, and man of God: "He could have added fortune to fame, but caring for neither, he found happiness and honor in being helpful to the world."[75]

---

[72] Carver Quotes. Posted on the George Washington Carver National Monument website at www.nps.gov/gwca

[73] Carver Quotes. Posted on the George Washington Carver National Monument website at www.nps.gov/gwca

[74] Dao, C. 2008. Man of Science, Man of God: George Washington Carver. Act & Facts. 37 (12): 8.

[75] Ibid.

A "Short" List of Peanut By-Products Discovered by G. W. Carver:

Peanut Punch	Flavoring Paste
Peanut Beverage Flakes	Meat Substitutes
All Purpose Cream (cosmetic)	Peanut Brittle
	Peanut Cake
Antiseptic Soap	Peanut Flour
Baby Massage Cream	Peanut Popcorn Bars
Face Bleach and Tan Remover	Peanut Relish
Facial Lotion	Peanut Tofu Sauce
Facial Powder	Salad Oil
Glycerine	Vinegar
Hand Lotion	Worcestershire Sauce
Peanut Oil Shampoo	Castor Oil Substitute
Shaving Cream	Emulsion for Bronchitis
Tetter and Dandruff Cure	Iron Tonic
Vanishing Cream	Laxatives
30 different Dyes for Cloth	Axle Grease
	Charcoal (from the shells)
19 different Dyes for Leather	Diesel Fuel
	Gasoline
17 different Wood Stains	Glue
Hen Food (from the peanut hearts)	Insecticide
	Linoleum
3 different kinds of Stock Food	Lubricating Oil
	Nitroglycerine
Bar Candy	White Paper (from the vines)
Caramel	Printer's Ink
Chili Sauce	Plastics
Chocolate Coated Peanuts	Rubber
Curds	Laundry Soap
Dry Coffee	Sweeping Compound

Image Source: Washington Carver, George. "A "short" List of Peanut By-Products Discovered by G.W. Carver." Image. Online *Encyclopedia Britannica*. The editors of *Encyclopedia Britannica*. https://www.britannica.com/summary/George-Washington-Carver, 17 April 2023.

ALBERT EINSTEIN

Albert Einstein was a German mathematician and physicist who developed the special and general theories of relativity. In 1921 he was awarded the Nobel Prize in physics for his explanation of the photoelectric effect. Within the next ten years he immigrated to the United States after he was targeted by the German Nazi Party.[76]

Einstein made major contributions to science including the following:

He provided empirical evidence for the atomic theory
He solved the riddle of the photoelectric effect
He proposed the special theory of relativity
He came up with the concept of rest energy through his famous equation
He proposed the general theory of relativity

---

[76] www.biography.com/scientist/Albert-enstein

He collaborated with Bose to predict the existence of
Bose-Einstein condensate
His debates with Bohr brought quantum mechanics
in focus
He was awarded the Nobel Prize in physics
His work had profound and far-reaching implications[77]

According to the BBA Foundation, Albert Einstein left his
mark on hundreds of technologies that we use every day such as
Google Maps, solar panels, self-lighting street lamps, and laser
beams.[78] All GPS navigators, including Google Maps, function
by measuring the distance from one point on Earth to one of
the many satellites that are currently orbiting our planet. The
time it takes for the signal sent by the satellites to reach the re-
ceptor is measured to calculate the distance. Einstein published
the theory of relativity in 1915. It explains why the clocks on
satellites are ahead of those on Earth by 38,000 nanoseconds.
That does not seem like much, but if these nanoseconds were
not accounted for, GPS systems would be off by kilometers.[79]

A year after Albert Einstein discovered general relativity, he
published his theory of stimulated emission, on which laser device technology is based. Today this technology is used in such
commonplace activities such as reading a bar code or using a
pointer in a presentation, but it is also used for delicate surgery
or industrial processes that require exact precision.[80]

---

[77] www.10majoracomplishments/Albert-enstein
[78] Four contributions Einstein has made to our daily lives (bbva.com)
[79] Ibid.
[80] Ibid.

In 1921 Einstein received the Nobel Prize for his discovery of the law of the photoelectric effect. That discovery made possible devices that turn light into electricity, like solar panels. That discovery also made way for automatic lighting of street lamps when night falls, the mechanism that stops elevators from closing when there is someone in the way, the device that regulates printer toner, and even breathalyzer tests.[81]

Did Einstein believe in God?

Albert Einstein's religious views have been the subject of much discussion. They have been widely studied and often misunderstood. He did not believe in a personal God who concerns himself with the fates and actions of humans. Einstein said that he was not an atheist. He also said that he believed in Spinoza's God.[82]

Baruch Spinoza was a philosopher of Portuguese-Jewish origin who lived from 1632–1677. He was born in Amsterdam, the Dutch Republic, and mostly known under the Latinized pen name Benedictus de Spinoza.

Among philosophers, Spinoza is best known for his *Ethics*, which presents an ethical vision unfolding out of a monistic metaphysic in which God and nature are identified. For Spinoza, God is no longer the transcendent Creator of the universe who rules it via providence, but nature itself, understood as an infinite, necessary, and fully deterministic system of which humans are a part.[83]

---

[81] Four contributions Einstein has made to our daily lives (bbva.com)
[82] en.wikipedia.org/wiki/Religious_and_philosophical_views_of_Albert_Einstein
[83] https://iep.utm.edu/spinoza/

In propositions one through fifteen of Part One, Spinoza presents basic elements of his picture of God. God is the infinite, necessarily existing (that is, self-caused), unique substance of the universe. He says that there is only one substance in the universe; it is God; and everything else that is, is in God.[84]

According to the information that I have gathered, in my opinion, neither Einstein or Spinoza were atheists, but both might be characterized as pantheists/humanists.

HUGH ROSS

Dr. Hugh Norman Ross is a Canadian astrophysicist, Christian apologist, and old-earth creationist.[85] Astrophysics is a branch of space science that applies the laws of physics and chemistry in an effort to understand the universe and our place in it. It

---

[84] "Baruch Spinoza," *The Stanford Encyclopedia of Philosophy* (Summer 2022 Edition), Edward N. Zalta (ed.), URL = <https://plato.stanford.edu/archives/sum2022/entries/spinoza/>.

[85] en.wikipedia.org/wiki/Hugh_Ross_(astrophysicist)

explores topics such as the birth, life, and death of stars, planets, galaxies, nebulae (according to NASA, a giant cloud of gas and dust in space), and other objects in the universe. The field of astrophysics is closely related to astronomy and cosmology.[86]

Christian apologetics is simply presenting a reasonable defense of the Christian faith and Christian truth to those who disagree.[87] Old-earth creationism is a belief that does not conflict with scientific evidence about the age of the earth.[88] Dr. Ross believes in progressive creationism, which is a view that while the Earth is billions of years old, life did not appear by natural forces alone, but that a supernatural agent formed different lifeforms in incremental (progressive) stages, and day-age creationism, a system of reconciling a literal Genesis account of creation with modern scientific theories of the age of the universe, the earth, life, and humans.[89]

Hugh Ross, PhD, is senior scholar and founder of Reasons to Believe. The organization is dedicated to communicating the compatibility of science and the Christian faith. RTB was established in 1986 by Dr. Ross, who was initially a skeptic, and later convinced by evidence that the words of the Bible are trustworthy. While he was in college, Hugh committed his life to Jesus Christ. He was convinced, after his study of cosmology, that there existed a Creator, specifically the God of the Bible.

[86] https://www.space.com/26218-astrophysics.html
[87] What is Christian apologetics? | GotQuestions.org
[88] https://en.wikipedia.org/wiki/Old_Earth_creationism
[89] en.wikipedia.org/wiki/Hugh_Ross_(astrophysicist)

Hugh Ross holds a degree in physics from the University of British Columbia and a PhD in astronomy from the University of Toronto. After five years on the Caltech faculty, (California Institute of Technology) he moved to full-time ministry and still serves on the pastoral team at Christ Church in Sierra Madre. He has written journal and magazine articles, hundreds of blogs, and a number of books including *The Creator and the Cosmos, Why the Universe Is the Way It Is, Improbable Planet, and Designed to the Core.* He has spoken on many university campuses, and at conferences and churches around the world.[90]

Ross was a former atheist who says that he grew up in a morally upright, but nonreligious home, and he was always surrounded by a community whose residents were mostly unreligious. Therefore, he had little exposure to Christians and Christianity during his youth. He said, "Our neighbors could also be described as nonreligious. I did not know any Christians or serious followers of any other religion while I was growing up."[91]

At the age of seven, Ross had developed a passion for physics. He was spending his time reading books on the sciences and physics. When he was sixteen, with his father's help, and the money he saved from collecting pop bottles, Ross was able to build his first telescope.

When he was just seventeen years old, Ross began serving as director of observations for Vancouver's Royal Astronomical

---

[90] Hugh Ross - Reasons to Believe
[91] How Science Led AstrophysicistDr. Hugh Ross To God | Reasons for Jesus

Society. At the same time, he began to study privately some of the world's sacred texts, testing them for accuracy.

Later, Ross would come to the conclusion that big bang cosmology presented a problem for his atheism. The next several years of his study convinced him that the universe had a beginning and, therefore, a *Beginner*, which was a Causal Agent outside or beyond the universe. But like the astronomers whose books he read, Ross imagined that the *Beginner* must be distant and noncommunicative.[92]

As Ross continued to study he began to find issues with the anti-religious perspectives and ideologies of the eighteenth century Enlightenment thinkers. Since Ross knew that the European philosophers of the Enlightenment largely discounted religion, his first inclination was to study their works. But what he discovered were inconsistencies, contradictions, evasions, and circular reasoning.[93]

But Dr. Ross compared the Bible with other religions, other sacred texts, and atheistic ideologies, and found the Bible to be "noticeably different." Therefore, it appealed to him as a scientist. He said,

> "It was simple, direct, and specific. I was amazed at the quantity of historical and scientific (i.e., testable) material it included and at the detail of this material. The first page of the Bible caught my attention. Not only

---

92 Ibid.
93 How Science Led AstrophysicistDr. Hugh Ross To God | Reasons for Jesus

did its author correctly describe the major events in the creation of life on Earth, but he placed those events in the scientifically correct order and properly identified the Earth's initial conditions."[94]

Does Hugh Ross believe in God?

For Ross, the Bible proved to be an exception, for it provided hundreds of statements that could be tested for accuracy, and it also anticipated, thousands of years in advance, many facts of sociopolitical history and of nature that research would only much later confirm. For example, Ross highlights the Bible's anticipation of the history and current tensions in the Middle East, as well as four fundamental features of big bang cosmology:

1. The beginning of space and time coincided with the beginning of matter and energy;
2. Continual expansion of the universe from the cosmic beginning
3. The constancy of physical laws; and
4. The pervasiveness of entropy (decay)

Ross, an astrophysicist, became convinced that belief in God was reasonable. The case so impressed him that he came to understand Jesus Christ as:

---

[94] Ibid.

"the Creator of the universe, that he paid the price that only a sinless person could pay for all my offenses against God, and that eternal life would be mine if I would receive His pardon and give him the rightful place of authority over my life...Through nearly two years of study, this book's predictive power persuaded me that it must have been inspired by One who knows and guides the past, present, and the future. I had essentially proven to myself that the Bible is more reliable than the laws of physics I focused on in my university courses. The only reasonable conclusion that I could see was that the Bible must be the inspired Word of God."[95]

Ironically, members on both sides of the debate do agree about one thing: big bang cosmology puts their position in jeopardy. The big bang poses a problem for young-earth creationists because it makes the universe billions of years old rather than thousands. Such an assertion undercuts their system at its foundation. Big bang cosmology also presents a problem for atheistic scientists because it points directly to the existence of a transcendent Creator—a fact they dare not concede.

---

[95] Ibid.

CHAPTER 5

Madame Curie, Florence Nightingale, and Katherine Johnson

Female scientists have made significant contributions throughout history. Among those brilliant women was Emilie du Châtelet (1706–1749) who studied mathematics and physics. She collaborated with the philosopher Voltaire, who also had a love of science. Her most lasting contribution to science was her translation of Isaac Newton's *Principia* into French. The *Principia* is still in use today.[96]

Ada Lovelace, mathematician (Dec. 10, 1815–Nov. 27, 1852) is respected as the first computer programmer. She recorded notes on Charles Babbage's proposed analytical engine

---

[96] Smithsonianmag.com

(a programmable, general-purpose computer) which is considered to be the first computer algorithm.[97]

MARIE CURIE

Marie Curie was born on November 7, 1867, in Warsaw, Poland, and died on July 4, 1934, in Passy, France. She was the youngest of five children. Madam Curie, is regarded as the first famous female scientist in the modern world. Because of her work in research on radioactivity she was known as the "Mother of Modern Physics." She coined the word after studying the radioactivity of radium in 1898.

Through her pioneering work on radioactivity, she earned two Nobel Prizes, and changed our basic understanding of radioactivity. Madam Curie championed the use of radiation in modern medicine. Her work opened the door to a new era of

[97] Ibid.

cancer treatment. Inspired by her research, compassion, and perseverance, generations of women in science and medicine have followed in her footsteps.[98]

FLORENCE NIGHTINGALE

Florence Nightingale was born on May 12, 1820, in Florence, Italy, and died on August 13, 1910, in London, England. A British nurse, statistician, and social reformer, Florence became known as the Lady with the Lamp because of her practice of entering soldiers' wards at night carrying a lantern and spending many hours caring for their physical and psychological needs. Florence was the foundational philosopher of modern nursing.[99]

---

[98] Britannica, The Editors of Encyclopaedia. "Marie Curie." Encyclopedia Britannica, 9 Oct. 2023, https://www.britannica.com/biography/Marie-Curie. Accessed 26 October 2023.

[99] 4 Selanders, Louise. "Florence Nightingale." Encyclopedia Britannica, 10 Oct. 2023, https://www.britannica.com/biography/Florence-Nightingale. Accessed 26 October 2023.

Although she was primarily remembered for her work during the Crimean war, Florence's greatest accomplishments focused on her attempts to create social reform in the area of health care and nursing. She improved the health of households through her most prominent publication, *Notes on Nursing: What It Is and What It Is Not*. That work provided direction on how to manage the sick. It has been in continuous publication worldwide since 1859.[100]

Other reforms were financed through the Nightingale Fund, and there was a school for the education of midwives established at King's College Hospital in 1862.[101] Nightingale improved the health of households through her most famous publication, *Notes on Nursing: What It Is and What It Is Not*, which provided directions on how to manage the sick. This volume has been in continuous publication worldwide, and is still available.

---

[100] Ibid.
[101] Ibid.

KATHERINE JOHNSON

Katherine Johnson was an African American mathematician who was born on August 26, 1918, in White Sulphur Springs, West Virginia, and died on February 24, 2020, in Newport News, Virginia.

At NASA Katherine was a member of the Space Task Group. She worked at NASA and its predecessor, the National Advisory Committee on Aeronautics for a total of thirty-three years.[102] In addition, Johnson played an important part in NASA's *Mercury* program (1961–1963) of crewed space flights. It was in 1961 that she calculated the path for *Freedom 7*, which was the spacecraft that put Alan B. Shephard Jr., the first U.S. astronaut, in space.[103] In 1962 Johnson verified, at the request of John Glenn, that the electronic computer had planned Shepherd's flight correctly. Consequently, John Glenn made history, becoming the first U.S. astronaut to orbit Earth.

---

[102] Wikipedia
[103] Ibid.

Katherine Johnson was a part of the team that calculated where and when to launch the rocket for the Apollo 2 mission of 1969, that sent the first three men to the moon.[104] On November 24, 2015, Katherine was awarded the Presidential Medal of Freedom by President Barack Obama at the White House in Washington, D.C.[105]

Katherine was a dedicated Presbyterian. She was a member of Carver Memorial Presbyterian Church in Newport News, Virginia. One of her several leadership roles in that church was its Finance Chair.[106]

---

[104] Wikipedia
[105] National Geographic.com/KatherinJohnson
[106] Presbyterianmission.org

CHAPTER 6

Benefits of Science

Have you thought lately about the benefits of science in your life today?

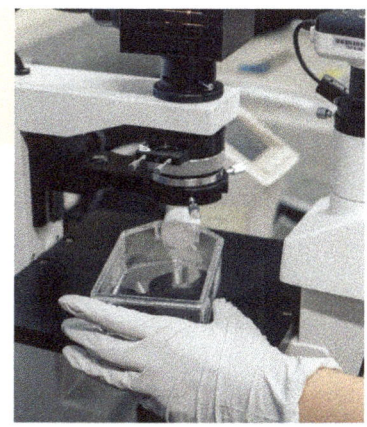

Photo credit: Jaron Nix, "a gloved hand preparing a stem cell culture for analysis," 04/25/18, image, The University of Alabama, Tuscaloosa, USA, 01/10/24, Unsplash.

It is a known fact that we benefit from the contributions of science every day, all day, and in every part of our lives. The alarm on my cell phone woke me up this morning at the precise time that I wanted to get up because I had asked it to do that. I literally spoke to the phone last night and ordered it to alarm at a particular time. When I first became an adult, I had to set my alarm by winding up my twin bell alarm clock. When I got up, I turned on the light, thanks to scientist Thomas Edison. That was quite a relief from the old kerosene lamp of my childhood. When I went to freshen up in preparation for breakfast, I felt especially thankful for modern plumbing facilities. It had been quite different growing up many years ago in rural Harford County, Maryland.

After breakfast I drove off to the YMCA for *Silver Sneakers*, a cardio program for seniors. Modern science made it possible to unlock my car from inside of my apartment, and to start it without putting a key in the ignition. In the event that I had forgotten the way to my destination I could have literally asked the car to remind me how to get there. My GPS would have gotten me there by the quickest route, and it would have warned me if there was slow traffic or an obstruction along the route. Throughout the day and into the evening, I took note of many of the benefits of science that we enjoy each and every day.

Many aspects of modern life are impacted by scientific discoveries.

Science fuels technology.

Because of basic science, we have modern devices such as automobiles, X-ray machines, computers, phones, and other communication devices. Basic science fuels technology, and technology affects our everyday lives. It also includes designed innovation, such as administering flu vaccines, the technique and tools used to perform open heart surgery, or even a new system of crop rotation. Simple things that might be easily taken for granted are science-based technologies: the plastic that makes up a sandwich bag, the genetically-modified canola oil in which fries are cooked, the ink in ballpoint pens, an ibuprofen—it's all here because of science.

Science helps in medicine.

A hundred years ago, a child affected by diabetes did not usually live more than a few years. However, because of the discovery of insulin in the early 1920s, along with subsequent scientific breakthroughs in genetic engineering, insulin can be mass-produced, and now diabetics live long lives.[107]

But diabetes is only one of many diseases and health concerns for which science has helped to develop treatments, preventions, and cures. Because of science we know how to build an X-ray machine, how to make an artificial knee, how to prevent nutritional deficiencies, how to prevent cholera and malaria. In fact, we wouldn't know that hand-washing can prevent the spread of germs if it weren't for science. In countless ways,

---

[107] Ibid.

science has supplied us with tools to improve human health—not the least of which has been medications to treat diseases.[108]

Science affects our personal decisions.

Although we may not be experts on microbiology, geology, or climatology, scientific knowledge may factor into our everyday decision-making. Science has implications for issues we face daily—and while science doesn't dictate which choice is the right one, it does give us important background knowledge to inform our decisions. Take hand-washing for example. A hundred and seventy years ago, hand-washing was not an everyday practice—even for doctors working in both the morgue and the maternity ward. But since then, biologists have developed the germ theory of disease, and research has shown that hand-washing prevents the spread of infection. A 2005 study found that promoting hand-washing among children in low-income areas could reduce the incidences of diseases like pneumonia by fifty percent.[109]

Dr. Ignaz Phillip Semmelweis lived from 1818 to 1865. He was a Hungarian physician and scientist who was an early advocate of antiseptic procedures. He was described as the "savior of mothers."[110]

Semmelweis found that, among women in the first section of the clinic, the death rate from childbed fever was two or three

---

[108] Ibid.
[109] https://undsci.berkeley.edu/understanding-science-101/what-has-science-done-for-you-lately
[110] Ignaz Semmelweis.-wiki en.m.wikipedia.org

times as high as among those in the second section, although the two sections were identical. There was one exception. Student interns were taught in the first, and midwives in the second. Semmelweis put forth the thesis that perhaps the students carried something to the patients they examined during labor. The death of a friend from a wound infection incurred during the examination of a woman who died of puerperal infection. Therefore the similarity of the findings in the two cases gave support to the reasoning. He concluded that the students who came directly from the dissecting room to the maternity ward carried the infection from mothers who had died of the disease to healthy mothers. He ordered the students to wash their hands in a solution of chlorinated lime before each examination. For the first time, deaths from childbed fever were dramatically reduced, from eighteen to one percent.[111] The leading cause of maternal mortality in Europe at that time was puerperal fever, an infection, which is now known to be caused by the streptococcus bacterium, that killed postpartum women.[112]

Semmelweis did not know the reason for his success, because the idea that germs existed and caused infections had not been thought of. In fact, the germ theory remained in obscurity for eighteen years until Louis Pasteur, a French biologist, microbiologist, and chemist, popularized it with his pasteurization method in 1865.[113]

---

[111] Ibid.

[112] https://theconversation.com/ignaz-semmelweis-the-doctor-who-discovered-the-disease-fighting-power-of-hand

[113] Ignaz Semmelweis: The Hungarian Doctor beaten to death for promoting Handwashing – HistoryVille (thehistoryville.com)

Dr. Semmelweis struggled for years in an effort to promote his ground-breaking contribution to the field of medical hygiene through his hand disinfection policies but was admitted to an asylum for mentally deranged people in 1865. While there, he was beaten and died fourteen days later, at the age of forty-seven, never having seen his life's work come to fruition.[114]

Science helps with vaccines and immunizations.

Vaccines teach our immune system to create antibodies, the same way it does when it is exposed to disease. But vaccines contain only killed or weakened forms of germs like viruses or bacteria. Therefore, they do not cause the disease, or put us at risk of its complications.[115]

Vaccines offer protection against many diseases, including: cervical cancer, cholera, COVID-19, diphtheria, hepatitis B, influenza, Japanese encephalitis, malaria, measles, meningitis, mumps, pertussis, pneumonia, polio, rabies, rotavirus, rubella, tetanus, typhoid, varicella, and yellow fever. Other vaccines are currently being piloted, including those that produce immunization against Ebola or malaria, but are not yet widely available in all countries.[116]

Science helps us in making food choices.

When we go to the fish market, what kind of fish do we choose? Say we have to decide between the local tilapia or the orange

---

[114] Ibid.
[115] https://www.who.int/health-topics/vaccines-and-immunization#tab=tab
[116] Ibid.

roughy. Taste factors into the decision, as does cost. But let's consider the role of science here. According to conservation biology, the orange roughy's population has been decimated by the seafood industry. Even more concerning, biologists have figured out that the fish lives to be one hundred years old, and does not begin to reproduce until it is twenty years old. That makes it difficult for the population to recover from over-fishing. Tilapia, however, is farmed specifically for human consumption and is not in danger of extinction. That helps us to make the best choice.[117]

Science helps us to choose where we want to live.

In the event that we are considering building a home or living in an area where earthquakes are frequent occurrences, we can consult with seismologists and geologists. They have taught us that not all types of soil are the same. Some areas within earthquake zones are unusually dangerous because of the possibility of liquefaction, which is a phenomenon where shaking causes soil particles to flow past one another easily, as in a liquid. In such a case, science could guide us to a safer area.[118]

Scientific knowledge also plays an important role in regulatory decision making and shaping public policy. A quart of milk is decorated with a nutrition label. Schools require and check students' vaccination records. Kitchen tiles are not made of asbestos. It is illegal to get rid of motor oil by pouring it

---

[117] https://undsci.berkeley.edu/understanding-science-101/what-has-science-done-for-you-lately
[118] Ibid.

down a storm drain. Science informs policies that help us to be safe, healthy, and good stewards of our environment.[119] I recognize that the preceding information provides an insufficient example of the impact that science has on our lives today.

---

[119] https://undsci.berkeley.edu/understanding-science-101/what-has-science-done-for-you-lately

CHAPTER 7

Some Benefits of Faith

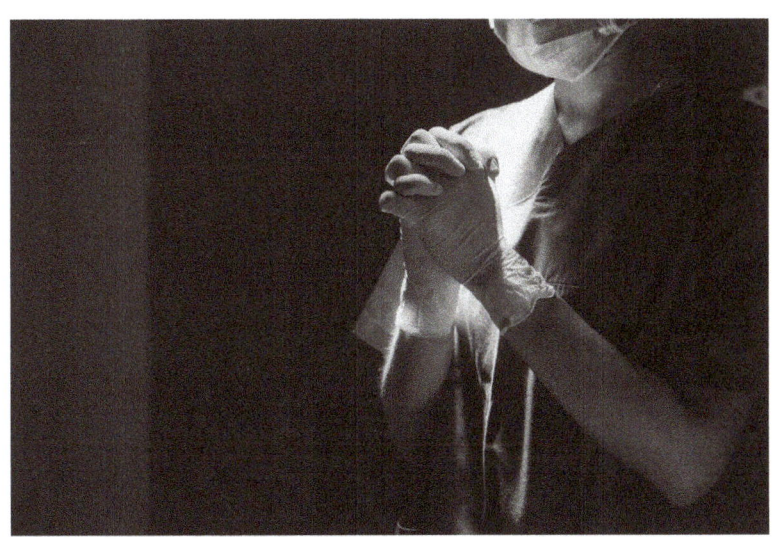

Photo credit: Getty Images Sven Hagolani.

"Now faith is the confidence in what we hope for, and the assurance about what we do not see."[120]

Faith gives us hope. We have faith in the present, and hope comes in the future as a result of faith. Faith brings hope into present-tense reality. Faith gives us support, determination, and power. When we feel overwhelmed by the challenges of life, faith helps us to overcome feelings of failure and defeat. Faith is an antidote for fear and anxiety.

Faith helps us to understand our purpose and our destiny. We receive confidence and a feeling of belonging from a community of faith. If we fall, faith can pick us up. If we fail, faith helps us to recover and try again. Faith provides for us a way of forgiveness. It helps us to persevere when circumstances seem to be impossible. We believe in God, we have prayer, and we believe in miracles. But we are not alone in our beliefs.

Oberammergau is the name of a village in Upper Bavaria, Germany, where a celebrated Passion play is performed by the villagers every ten years.

A little more than twenty years ago, my wife Margie and I went on a tour, sponsored by Educational Opportunities, a travel company operating out of Florida. Included in the tour were Austria, Belgium, Germany, and Switzerland.

One of the most memorable highlights of our trip was the Passion play in Oberammergau. Before we went on the tour, I did not recall ever having heard about either the town or the drama. In order to help the reader to understand the role

---

[120] Hebrews 1:1 NIV Bible

of faith in this story, I have provided what I hope is adequate historical background.

In the mid-1300s a calamitous epidemic struck Europe and Asia. It was a bubonic plague called the Black Death. It was in October of 1347 when twelve ships from the Black Sea docked at the Sicilian port of Messina. People who had gathered on the docks were met with terrifying surprise: most of the sailors aboard the ships were dead, and those still alive were extremely ill and covered in black boils that dripped blood and pus. Sicilian authorities quickly ordered the fleet of ships out of the harbor. But it was too late! Over the next five years the bubonic plague, the Black Death, would kill more than twenty million people in Europe which was almost one-third of the continent's population.[121]

There had been rumors among many Europeans about a "great pestilence" that was leaving a path of death across the trade routes of the Near and Far East. As a matter of fact, in the early 1340s the plague had struck China, India, Persia, Syria, and Egypt.[122]

Where did the Black Death come from?

What were its symptoms, and how did it spread? Scientists don't really know where the plague originated. It is thought to have originated somewhere in Asia more than 2,000 years ago, and spread by trading ships. However, recent research has

---

[121] Bubonic Plague (Black Death): What Is It, Symptoms, Treatment (clevelandclinic.org)
[122] Ibid.

indicated that the pathogen (germ) responsible for the Black Death may have existed in Europe as early as 3,000 BC.[123]

According to Cleveland Clinic, a plague is an infectious disease that is caused by a specific type of bacterium called *Yersinia pestis*. Y. pestis can affect humans and animals and is spread mainly by fleas. Bubonic plague is one type of plague that gets its name from the swollen lymph nodes (buboes) caused by the disease. The nodes in the armpit, groin, and neck may become as large as eggs and may ooze pus. There are other types of plagues: septicemic plague, which occurs when the infection goes all through the body, and pneumonic plague, that happens when lungs are infected. This is the same bubonic plague that killed over twenty-five million people during the fourteenth century. Rats traveled on ships and fleas and plague came with them. The bubonic plague was the Black Death because the people who got the plague died, and many of them had blackened tissue, caused by gangrene. A cure for the plague was not available at that time.[124]

Does the plague exist today?

What are the symptoms? How is it diagnosed and treated? When the bubonic plague first devastated Europe, physicians relied on crude and unsophisticated techniques such as bloodletting and boil-lancing. Those practices were both dangerous and unsanitary. Treatments also included superstitious practices

---

[123] Ibid.
[124] Bubonic Plague (Black Death): What Is It, Symptoms, Treatment (clevelandclinic.org)

such as burning aromatic herbs and bathing in rosewater or vinegar.[125]

According to the Cleveland Clinic, there have been other episodes of bubonic plague in the world apart from the Black Death years (1346–1353). Even today bubonic plague occurs throughout the world and in the U.S., with cases in Africa, Asia, South America, and the western areas of North America. About seven cases of plague happen in the U.S. every year, with half of those cases involving people aged twelve to forty-five. In the U.S., most cases are found in the following areas: northern New Mexico, northern Arizona, and southern Colorado, also, southern Oregon and western Nevada.[126]

Symptoms of bubonic plague include: sudden high fever and chills, pains in the areas of the abdomen, arms and legs, headaches, large swollen lumps in the lymph nodes that develop and leak pus. Symptoms of the septicemic plague might include blackened tissues from gangrene, often involving the fingers or toes, or unusual bleeding. People who have pneumonic plague may have problems breathing, and may cough up blood, and have nausea and vomiting. Plague can be diagnosed through tissue samples or blood tests.[127]

Since plague has not yet been eradicated, I thought it prudent to add a word of caution. Bubonic plague is a type of infection that is caused mostly by fleas on rodents and other animals. If humans are bitten by fleas, they can become infected.

---

[125] Black Death - Causes, Symptoms & Impact | HISTORY
[126] Bubonic Plague (Black Death): What Is It, Symptoms, Treatment (clevelandclinic.org)
[127] Ibid.

Cats can be infected by eating sick rodents, and those cats can pass droplets infected with plague to their owners or to the veterinarians who treat them.[128] But thanks to modern science, the plague can be treated and cured. Antibiotics that can be used to treat and cure bubonic plague include: Ciprofloxacin, levofloxacin and moxifloxacin, Gentamicin and Doxycycline.[129]

Oberammergau is a picturesque village that is located in the Bavarian Alps, on the Ammer River. It is a quiet, clean, and beautiful town. The houses remind you of a land of fairy tales. It has a population of only 5,125, according to their latest census. When we first arrived in the village of Oberammergau, we knew very little about its history.

Before the drama began, the narrator told us that the play had been performed every ten years from 1634 to 1674, and every ten years since 1680. He informed us that every participant must have lived in the village for at least twenty years, and that all beards were required to be natural. The play lasts for eight hours, and is performed under the open skies without protection against the weather.

According to the narrator, there was an outbreak of bubonic plague that devastated Bavaria during the Thirty Years' War, which lasted from 1618 to 1648. The village of Oberammergau remained plague-free until September 25, 1633. There was a man named Kaspar Schisler who had been working in another village. Those workers were forbidden to return home during the plague. But Kasper came home during the night

---

[128] Ibid.
[129] Bubonic Plague (Black Death): What Is It, Symptoms, Treatment (clevelandclinic.org)

to see his wife, and by accident, brought the plague with him. The man reportedly died, and in the following thirty-three days eighty-one villagers died from the plague. That was half of Oberammergau's population. On October 28, 1633, the villagers gathered in the church to pray. There they made a vow that if God would spare them from the plague, they would perform a play every ten years, depicting the life and death of Jesus, until the end of time. The narrator told us, "nobody died of plague in Oberammergau after that vow, and all of the town members that were still suffering from the plague recovered. The villagers kept their word to God by performing the play for the first time in 1634."

The drama is now performed in years ending with a zero, as well as in 1934, which was the 300th anniversary, and in 1984, the 350th anniversary. The 1920 performance, however, was cancelled due to the beginning of the Second World War in 1939. The play involves more than 2,000 actors, singers, instrumentalists, and technicians. All participants are residents of the village. The 2020 performance was cancelled due to the outbreak of COVID-19.[130]

The next performance of the Passion Play is scheduled for the year 2030 when nearly half a million visitors are expected.[131]

Our Lady of Lourdes is the title of the Virgin Mary. She is venerated under this title by the Roman Catholic church due to her apparitions that the church claims have occurred in Lourdes, France.[132]

---

[130] Oberammergau Passion Play - Wikipedia
[131] www.europeantraveler.com
[132] The Story of St. Bernadette and Our Lady of Lourdes - The Catholic Company®

St. Bernadette of Lourdes, a French nun who lived in the 1800s, claimed to have had visions of the Virgin Mary when the nun was only a teenager. Those experiences ultimately led to the founding of the shrine of Lourdes.

In the face of strong opposition, St. Bernadette staunchly defended her visions of Mary, and Pope Pius IX authenticated those visions in 1862. Since Mary has been venerated as Our Lady of Lourdes, the shrine of Lourdes has become a major site of Roman Catholic pilgrimage.[133]

Bill Whitaker, an American journalist working as a correspondent for CBS News's *60 Minutes*, has covered major news stories in the United States and around the world. On December 22, 2022, Mr. Whitaker reported on a very inspiring miracle story; that of Sister Bernadette Moriau. According to Mr. Whitaker, the Sanctuary of Our Lady of Lourdes in Southern France is the location where seventy medical miracles have taken place, as authenticated by the Catholic Church.

In his report in December of 2022, Bill Whitaker said that the Marian shrine was famous to the faithful, but that the Lourdes Office of Medical Observations was well known. He said,

"That's where world-renowned doctors and researchers conduct decade-long investigations into the countless claims of cures reported over the years. They determine which cases can be medically explained and which

[133] Ibid.

cannot—it's what those church officials might call a miracle. For the doctors, it's a lesson in the limits of medicine. For the devout, it's a divine intervention."

For his report, Bill Whitaker interviewed Sister Bernadette Moriau, visited the Marian Shrine, and the Lourdes Office of Medical Observations.[134]

Sister Moriau was eighty-three years old at the time of the interview with the *60 Minutes* reporter. When Sister Bernadette had traveled to Lourdes fourteen years ago, she did not know that she would be healed. She told Whitaker that her physician had persuaded her to take a pilgrimage to Our Lady of Lourdes in 2008. At that time she was suffering from cauda equina syndrome, which is a debilitating condition that occurs when the bundle of nerves below the spinal cord are damaged. Reportedly in "full, total paralysis," Moriau arrived at Lourdes in a wheelchair.

During a walk in Bresles, France, Sister Bernadette told Bill Whitaker that her left foot was once crippled, and that in order to walk she had to have a back and leg brace, and an implant to dull nerve pain as well as exceptionally large doses of morphine. But she said that her condition changed when she went to Lourdes.[135]

She also pointed out the fact that she did not go to Lourdes to receive a miracle, and that although she believed in miracles,

---

[134] https://www.cbsnews.com/news/sanctuary-of-our-lady-of-lourdes-miracles-cures-2022-12-18/
[135] Ibid.

she was not expecting to be healed. She just planned to go there and pray for others. It was after she joined in procession with others that she received her miracle.

Sister Moriau returned home feeling spiritually strengthened and uplifted, but physically, she felt worse. After spending three days in exceedingly intense pain, she suddenly felt strong enough to walk to the chapel to pray. She told correspondent Whitaker,

> "Then I felt some kind of heat coming into my body. I felt relaxed, but I didn't really know what that was meaning. And in my room, I heard the inner voice again telling me, 'Take all your braces off.' I didn't think twice. And I started taking my foot brace off. And my foot that used to be crooked was straight. And I could actually put it on the ground without feeling any pain."[136]

Dr. Alessandro de Franciscis has been a practicing physician for more than thirty years. He is a former pediatrician and Harvard-trained epidemiologist. But he refers to himself as a "useless doctor." In 2009 Dr. de Francisis was appointed by the bishop of Tarbes to be the fifteenth *Médecins Permanent*—the president of Bureau des Constatations Medicales de Lourdes, which is the Lourdes Office of Medical Observations. The office was founded in 1883 for the purpose of recording, studying, and judging hundreds of cures that were reported by pilgrims

---

[136] Ibid.

who came to Lourdes to wash in the waters of the spring, revealed by Mary (mother of God) when she appeared to St. Bernadette Soubirous. Since the office was founded, 7,000 cases of unexplained cures from grave medical conditions have been recorded. Only a total of sixty-nine of those investigated by the Lourdes Office of Medical Observations had been accepted as miracles by church authorities. Dr. de Francisco jokingly calls himself a "useless doctor" because his primary responsibility is to evaluate patients who have already been cured![137]

When he was speaking to Bill Whitaker about Sister Bernadette Moreau he said, "We sent her to different neurologists because of the different specific case of her disease. We asked to repeat twice some imagery—electrophysiology." He said, "We did all that you would do in medicine to make absolutely sure of her [diagnosis]. And it was," he said implying that it was a miracle. On top of that we saw two psychiatrists in Paris.[138]

Subsequently, Dr. de Franciscis sent Sister Bernadette's case to a group of thirty-three doctors and professors; the International Medical Commission of Lourdes to determine whether her cure was "medically unexplained." Their conclusion was that it was, after eight years of investigation. Dr. de Franciscis told Bill Whitaker, "If tomorrow morning any of our viewers is a doctor, and one day he stops in southern France and comes to see me and wants to look into the file of Sister Bernadette, I'll be delighted to show him."[139]

---

[137] The "useless" doctor of Lourdes – Catholic World Report
[138] Ibid.
[139] Ibid.

CHAPTER 8

Is There a Convergence?

I believe in science, and I believe in God.

Photo Credit: Wikipedia. 2024. "Caduceus" Wikimedia Foundation.

Photo Credit: James, "cross," 10/03/17, imag, 01/10/24, Unsplash.

I confess that my knowledge of science is quite limited; my knowledge of God is limited as well. My limited knowledge of science, however, does not, in any way, limit the possible benefits of science for me or for the world I live in. Likewise, my limited knowledge of God does not, in any way, limit God's power or possibilities for me or for the world. Denial of facts does not change the facts. People still deny the Holocaust, the Armenian genocide, and other historical facts. Still others deny that we sent people to the moon in the 1960s, despite a plethora of evidence and documentation.

It is a fact that faith and science are compatible. Faith and science do coexist, from the cosmic scale of gravity down to the micro scale of the atom, faith undergirds scientific knowledge.[140] There are people of many different faith backgrounds, and many levels of scientific knowledge see no contradictions between science and religion. Many people simply admit that the two institutions deal with different realms of human experience.[141]

Science is humankind's endeavor to comprehend the way that the world functions. One of the greatest instruments available to accomplish that task is the scientific method.[142]

It starts with a question about the world. Then background research, an experiment, analysis to determine if the hypothesis was correct, the cycle is finished. If not, another hypothesis is put forth, and testing begins again. The scientific method

---

[140] www.allaboutscience.org/science-and-faith.htm
[141] www.allaboutscience.org/science-and-faith.htm
[142] Ibid.

infers that a provable fact will be repeatable and verifiable—that other scientists will come up with the same answer if their experiment is performed in the same way.[143]

The scientific method is not in conflict with biblical teaching, essentially. God created light matter, vegetation, animals, and human beings. As we endeavor to comprehend God's amazing creation, we give Him honor. We continue to grow in the knowledge of God, of his wisdom and power and majesty. We value more deeply God's unmerited favor as we come to understand the meanings of His miracles. Seeing the cancerous tumor disappear from an MRI after seeing it in the previous one causes us to be increasingly grateful.[144]

Since God exists, and is the Creator of the universe, faith in God and belief in science do not contradict. If God did not exist, then faith and science would indeed be at odds with each other, since scientists search for facts about the cosmos which God created.[145] It is clear that science and faith were meant to work together, and that true science will guide us into a deeper understanding of our Creator. It will reveal to us ways that we can better serve one another. We need both faith and science, unfettered from those who pervert and misuse them.[146]

---

[143] Ibid.

[144] Htps://.compellingtruth.org/faith-vs-science.html146

[145] Htps://.compellingtruth.org/faith-vs-science.html146

[146] sojo.net/articles/how-faith-and-science-work-together

CHAPTER 9

What Science and Faith Have Done for Me Lately

I knew something was wrong.

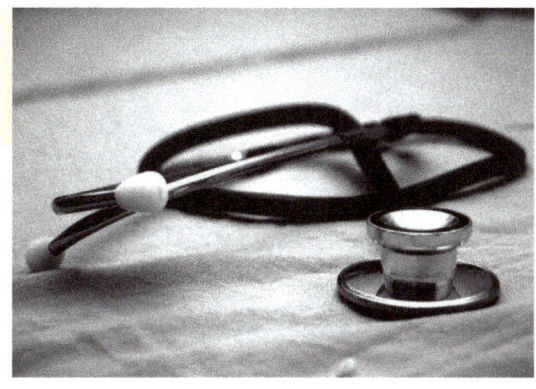

Photo credit: Hush Naidoo Jade Photography, "Stethoscope," 09/18/2017, image, Johannesburg, South Africa, 01/10/24, Unsplash.

At first I thought my shortness of breath after walking up a flight of steps or walking for a couple of blocks was only an indication that my age was catching up with me. Also, my family doctor had told me that I should lose a few pounds. It was not until I was admitted to Lancaster General Hospital that I learned of my heart attack.

Early one afternoon, following a hearty brunch with some friends, I began experiencing symptoms of what I thought was food poisoning, accompanied by excruciating pains in my back and left shoulder. I took some acetaminophen and went to bed. The medicine eased the pain, but early the next morning the pain in my back returned. I called for an ambulance and was transported to the hospital shortly thereafter.

Preliminary diagnostic tests revealed that I had experienced a heart attack. I was scheduled for a cardiac catheterization. A cardiac catheterization is a procedure used to diagnose and treat certain heart conditions. A doctor inserts a long, thin tube catheter through a blood vessel and into the heart. The procedure can find out the cause of chest pains, abnormal heart rhythm, valve disease, or coronary artery disease.[147] It can also evaluate heart muscle function, take biopsies, or check pulmonary arteries.[148]

The catheterization revealed that I had had a minor heart attack, and that there were a couple of other people who got scheduled before me because their conditions called for immediate intervention. I spent more than a week in the hospital,

[147] www.mayoclinic.org/tests-procedures/cardiac-cath
[148] Ibid.

waiting to be scheduled for the catheterization, which revealed that in order to correct the problems in my heart, I would need what the doctor referred to as considerable plumbing work. I needed triple bypass heart surgery.

As I think about my longest and most serious hospitalization, I realize that it is incumbent on me to express my gratitude for faithful friends who have prayed for me, the women and men who have cared for me during my stay at Lancaster General, and the surgical team that operated with great knowledge, precision, and compassion.

Prior to either of my medical procedures at Lancaster General I had a visit from one of my healthcare providers, an MD, a specialist who has been in private practice in the Lancaster area for over thirty years. I was greatly comforted as he and his wife offered prayers for my successful treatment and recovery. I also deeply appreciated the pastoral visit and personal prayers by my district superintendent, the Reverend Jennifer Freymoyer.

In addition, prayers from faithful family members, church members, and other friends were offered up on my behalf. I have felt the effects of those prayers of faith that have most assuredly been answered. They have helped me to hold onto hope that God would be with me throughout the entire experience and that I would emerge healthier and stronger than before. According to a study by the George Washington Institute for Spirituality and Health, after oral pain medication, prayer is the second most common method of pain management, and the most common method of pain management without using drugs. The study found that prayer could reduce feelings of isolation, anxiety, and fear, which could lead to lower blood pressure, and improved immune functioning.[149]

During the time that I was an inpatient I had ample time to pray and to think about my faith. Faith has truly done a lot for me lately!

I appreciated the holistic nursing care that I received at LGH. By holistic nursing care I refer to care of the mind, body, and soul. They helped me to heal in various ways. They were available when I needed them. The women and men who worked as nurses in the cardiology unit were intuitive. They seemed to know what I needed before I told them. It was clear that they were dedicated and enjoyed their work. Although they were responsible for the care of many patients, they seemed to do all in their power to make me feel that I was their only patient. The clinical technicians were there to assist the nurses in caring for the patients in any way they could. I deeply appreciated the personal care they were willing to provide for me without the slightest reluctance. They, too, were skilled, and took pride in their work. I do not remember the names of the staff members, but I cannot forget how important they made me feel!

Science has done a lot for me lately. Case in point, in August of 2023, I had coronary artery bypass graft surgery. Before the day of my surgery my doctor explained the whole process to

---

[149] https://www.takingcharge.csh.umn.edu/prayer

me. He also said that it was a question of plumbing that needed to be done in my heart to restore adequate blood flow.

The area of surgery was marked in order to make sure that there would be no errors.

I was kept comfortable and safe by my anesthesiologist. I was asleep and unaware of the process during the three or so hours of the operation.

The doctor made the bypass with a healthy piece of blood vessel, harvested from my right leg. He attached the healthy blood vessel to the blocked artery. The new blood vessel bypassed the diseased artery in order to increase blood flow to my heart muscle. The surgeon made an incision in my sternum (breast bone). He then cut through my sternum in order to reach my heart and coronary arteries. He connected me to a heart-lung bypass machine. It added oxygen to my body and moved the blood through my body. That machine permitted the doctor to stop my heartbeat while he was working on my arteries. When the blood vessels were in place, he restarted my heart. During the surgical process, my doctor also performed a cardiac ablation, using heat to create tiny scars in my heart to block irregular electrical signals and restore a typical heartbeat. The procedure was used to correct heart rhythm problems known as atrial fibrillation.

After surgery, I was moved to the intensive care unit, where I stayed for two days. Then I went back to the cardiology unit. I was on a liquid diet for a day or so, and I had many tubes and wires attached to my chest, arm, neck, et cetera.

Within a week from the day of my surgery, I was sent home to continue my recovery. There I received excellent follow-up care, which included frequent visits from visiting nurses, an occupational therapist, a physical therapist and a social worker.

After I was dismissed by those therapists, I was assigned to cardiac rehabilitation therapy for thirty-six sessions. I was also urged to exercise regularly, and to eat a healthy diet, avoiding foods that are high in sodium and saturated fat.

Science has done a lot for me recently!

Photo credit: Barbara Olsen, "Unrecognizable black man training in park," 2023, image, 01/10/24, Pexels.

CHAPTER 10

Why the Bible and Christianity

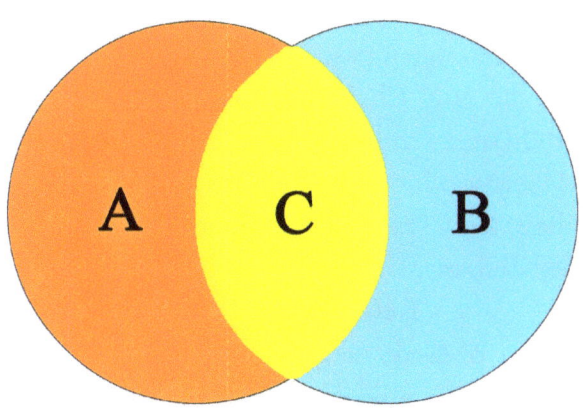

Photo Credit - Author

According to Encyclopedia Britannica, the Bible, which contains the sacred writings of Judaism and Christianity, has been, for us in the West, a great resource. It has been our most accessible and trustworthy resource and authority on intellectual, moral, and spiritual ideals.[150]

Hebrew was one of the original languages of the Bible. It was translated into Greek, the Septuagint, during the Hellenistic period. The Septuagint was written in Alexandria because Alexander the Great had conquered many surrounding nations, and Greek had become the predominant language. The original law was translated into Greek for those Jews who no longer spoke Hebrew, and also to convert many Greeks to Judaism.[151] The Septuagint became known as the Greek Old Testament. It also provided an original language for the New Testament for the first three centuries CE. The New Testament often uses the Septuagint when referring to the Old Testament. Paul often made use of the Septuagint when he wanted to share an Old Testament passage with people who spoke Greek. Leaders of the early Christian church, such as the Apostolic Fathers, and Greek Church Fathers did the same.[152]

The Hebrew Bible is divided into three sections: the Law, the Prophets, and Writings. For Christians, those books make up the Old Testament. The Bible focuses on the one and only God, the Creator of everything. The Old Testament begins with the story of God's creation of the world and tells of the Israelites and the Promised Land. The New Testament gives an

---

[150] Britannica, The Editors of Encyclopaedia. "Bible." *Encyclopedia Britannica*, 27 Mar. 2023, https://www.britannica.com/topic/Bible. Accessed 17 May 2023.
[151] Ecclesia.org/truth/Septuagint.html
[152] overviewbible.com/Septuagint/

account of the life, the person, work, and teachings of Jesus, and the genesis and development of the Christian Church.[153]

Christianity as a religion came from Judaism, but as a community of faith, it owes its existence to Jesus Christ. After several centuries of development through lawgivers, priests, and prophets, Judaism became the major religion of Israel. Religions and political conflicts had caused a division in the nation of Israel into two distinct kingdoms, one to the north and the other to the south. Both kingdoms were conquered by aggressors from the Tigris-Euphrates valley. In the meantime, many of the people, including many important leaders, were deported, and only a small group, a remnant remained.[154]

It was through the good will of Persian kings, like Cyrus, that Palestine became included within their boundaries, and a new temple was built in Jerusalem. The temple became the center for the remnant as well as those who had been scattered in the dispersion.

The term *Jew* refers to a person whose religion is Judaism. In broader terms a Jew is a member of a worldwide cultural group that traditionally practices the Jewish religion. The word itself comes from the Hebrew word *Yehudi*, the Latin *Judaeus*, and originally meant "belonging to the tribe of Judah." Judah was one of the twelve tribes of Israel, descended from Judah who was the fourth son of Jacob.[155]

---

[153] Britannica, The Editors of Encyclopaedia. "Bible." *Encyclopedia Britannica*, 27 Mar. 2023, https://www.britannica.com/topic/Bible. Accessed 17 May 2023.

[154] www.britannica.com › topic › Christianity

[155] Genesis 29:35, New Living Translation

After their escape from Egypt, the tribe of Judah entered Canaan and settled in the region south of Jerusalem. In time Judah became the strongest and most influential tribe, producing two great kings: David and his son Solomon. Very early in history, the Hebrew prophets predicted the coming of Christ. It was predicted and expected that a Messiah, a promised deliverer, a son of David would come to help the Jews shake off the yoke of Roman oppression.

The Messiah in Judaism is a future Jewish king from the line of David who is expected to be anointed with holy oil and rule the Jewish people during the Messianic age and world to come. He is a savior and liberator who is believed to be the future redeemer of the Jewish people.[156]

Christians believe that Jesus is the Messiah who came to deliver people from the power and penalty of sin. The gospel proclaims that Jesus paid sin's penalty through His substitutionary death for sin. According to the Christian understanding of the Bible, Jesus died, was buried, and rose again, proving that God accepted His payment for sin. Jesus was punished for our sins so that we could be forgiven and avoid sin's punishment. This is the Christian understanding of true salvation.

The Gospel of Jesus Christ is Good News, and the heart of that Gospel is John 3:16, undoubtedly the most well-known verse in the Bible. "*For God so loved the world that He gave His only begotten Son, that whoever believes in Him should not perish*

---

[156] https//en.Mwikipedia.org

but have everlasting life."[157] I like to include in my sermons that verse seventeen is also very important in understanding Jesus and his mission. It says, "*For God did not send his Son into the world to condemn the world, but that the world, through Him, might be saved.*"

We deeply appreciate the time and attention that you have devoted to reading this book. We hope it has informed, enlightened, encouraged, and inspired you in some way or other. We realize that a book is not complete just because it has been written. It must be read. Thank you!

---

[157] John 3:16, New King James Version

ACKNOWLEDGEMENTS

I am grateful to my brother, David, for his encouragement, guidance in the areas where science and faith connect, practical application of scientific principles, and financial backing for production and marketing, without which this book would not have been published.

READER'S NOTES

READER'S NOTES

_____

_____

_____

_____

_____

_____

_____

_____

_____

_____

_____

_____

ABOUT THE AUTHORS

Dr. Garfield Greene

Dr. Garfield Greene, honorably discharged in 1961, served as an American Airman in the United States Air Force. He has a BA in French from Morgan State College, an MSW from the University of MD at Baltimore, an M. Div. from Princeton Theological Seminary, and a Doctor of Ministry from Wesley Seminary in Washington, D.C.

He has been employed as a Chaplain at the Trenton State Prison, a Clinical Social Worker for the Dept. of the Army, and for the V.A., and has served as a United Methodist Pastor in both New Jersey and Pennsylvania.

He has published two other books, *Ordinary People* and *What Do You Say?*.

David E. Greene

David E. Greene is currently the Owner and Administrator of Medicaid Personal Providers, LLC, a State of Maryland licensed residential service agency since 2002. In 2016, David E Greene led Medicaid Personal Providers, LLC's successful efforts for the Joint Commision Accreditation for Homecare Organizations (JCAHO).

He is a former member of the White House Press Corps under President Jimmy Carter and under President Ronald Reagan. He is a former consultant to the combined auto industry developing countermeasures to curb the use of illegal drugs in the auto production process.

He served as the Vice President of the International Policy Institute in Washington, D.C., collaborating closely with Professor Harley Henrichs, who has since retired. Together, they organized political-economic seminars tailored for ambassadors from the 212 embassies situated in D.C. David E. Greene also maintained a Top Secret clearance under Presearch Corporation of Silver Spring in 1975 and was a co-author of the Mine Warfare Countermeasures Handbook.

David E. Greene graduated from the South Eastern University of Washington, DC with a Masters of Business Administration and subsequently a Masters of Public Administration.